Environmental Chemistry at a Glance

Ian Pulford

Senior lecturer, Department of Chemistry, University of Glasgow

Hugh Flowers

Lecturer, Department of Chemistry, University of Glasgow

Blackwell
Publishing

© 2006 by Ian Pulford and Hugh Flowers

Blackwell Publishing Ltd Editorial offices:
Blackwell Publishing Ltd, 9600 Garsington Road, Oxford OX4 2DQ, UK
  Tel:+44 (0)1865 776868
Blackwell Publishing Inc., 350 Main Street, Malden, MA 02148-5020, USA
  Tel:+1 781 388 8250
Blackwell Publishing Asia Pty Ltd, 550 Swanston Street, Carlton, Victoria 3053, Australia
  Tel:+61 (0)3 8359 1011

The right of the Author to be identified as the Author of this Work has been asserted in accordance with the
Copyright, Designs and Patents Act 1988.

All rights reserved. No part of this publication may be reproduced, stored in a retrieval system, or transmitted, in
any form or by any means, electronic, mechanical, photocopying, recording or otherwise, except as permitted by
the UK Copyright, Designs and Patents Act 1988, without the prior permission of the publisher.

First published 2006 by Blackwell Publishing Ltd

ISBN-10: 1-4051-3532-8
ISBN-13: 978-1-4051-3532-0

Library of Congress Cataloging-in-Publication Data

  Pulford, I. (Ian)
  Environmental chemistry at a glance / I. Pulford, H. Flowers.– 1st ed.
    p.   cm.
  Includes index.
  ISBN-13: 978-1-4051-3532-0 (pbk. : alk. paper)
  ISBN-10: 1-4051-3532-8 (pbk. : alk. paper)   1. Environmental
  chemistry.   2. Chemistry, Inorganic.   I. Flowers, H. (Hugh)   II. Title. QD31.3.P85 2006

  577$'$.14 – dc22
  2006002455

A catalogue record for this title is available from the British Library

Set in 9.5/11.5 pt Times by TechBooks, New Delhi, India
Printed and bound in Great Britain
by TJ International Ltd, Padstow, Cornwall

The publisher's policy is to use permanent paper from mills that operate a sustainable forestry policy, and which
has been manufactured from pulp processed using acid-free and elementary chlorine-free practices. Furthermore,
the publisher ensures that the text paper and cover board used have met acceptable environmental accreditation
standards.

For further information on Blackwell Publishing, visit our website:
www.blackwellpublishing.com

Contents

A	**Chemistry of the Surface Environment**	1
1	Rocks and Minerals	2
2	Weathering Processes	4
3	Clay Minerals	6
4	Hydrous Oxides	8
5	Humified Organic Matter	10
6	Flocculation and Dispersion of Clays, Oxides and Organic Matter	12
7	Ion Exchange	14
8	Adsorption	16
9	Solubility Processes	18
10	Complexation and Chelation by Organic Matter	20
11	pH and Buffering	22
12	Redox Potential	24
B	**Soil**	27
13	Soil Development	28
14	Soil Horizons and Soil Profiles	30
15	Common Soil Types	32
16	Soil Texture	34
17	Soil Structure and Aggregation	36
C	**Sediments**	39
18	Sediments – Processes	40
19	Sedimentary Environments	42
D	**Water**	45
20	The Hydrological Cycle	46
21	The Freshwater Environment	48
22	The Marine Environment	50
E	**Atmosphere**	55
23	The Atmosphere	56
24	Atmospheric Processes	58
F	**Biosphere**	61
25	Nutrient and Energy Flows in the Biosphere	62
26	Diversity of Micro-organisms	64
27	Inorganic Plant Nutrients	66
G	**Chemical, Physical and Biological Interaction**	69
28	Temperature of Environmental Systems	70
29	Soil Water	72
30	Aeration and Gas Exchange	74
H	**Environmental Cycles**	77
31	The Carbon Cycle	78
32	The Nitrogen Cycle	80
33	The Sulfur Cycle	82
34	The Phosphorus Cycle	84
35	The Potassium Cycle	86

Contents

I Pollution — 89
- 36 Sources of Pollution — 90
- 37 Pesticides — 92
- 38 Fertilisers in Agriculture — 94
- 39 Nitrate Leaching — 96
- 40 Nitrate in Drinking Water — 98
- 41 Sewage Treatment — 100
- 42 Environmental Impacts of Sewage — 102
- 43 Contaminated Land — 104
- 44 Organic Contaminants in the Environment — 106
- 45 Heavy Metals — 110
- 46 Environmental Impacts of Mining — 114
- 47 Radioactivity — 116
- 48 Environmental Impacts of Radioactivity — 118
- 49 Global Warming and Climate Change — 120
- 50 The Ozone Layer — 122
- 51 Damage to the Ozone Layer — 124
- 52 Acid Rain — 128

Index — 131

Chemistry of the Surface Environment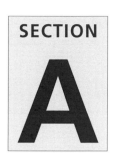

SECTION A

1. Rocks and Minerals

Weathering processes act upon rocks of the Earth's crust which are exposed at the surface, bringing about changes in their make-up and characteristics. These processes can be physical, chemical or biological, and the products of weathering can be new minerals, altered minerals or soluble components that are released into the wider environment.

The predominant mineral types in these rocks are silicates, which also dominate the characteristics of most soils. In Table 1.1 elements in the crust with an average concentration of greater than 1% are listed, along with their corresponding soil content. The enrichment factors of approximately 1 for O, Si, and Al demonstrate the importance of aluminosilicates in soil. Values <1 show loss of these elements from soil as a result of weathering processes.

Table 1.1 Average elemental concentrations in the Earth's crust and in soil, and the enrichment factor in soil (soil:crust ratio).

Element	Average concentration in crust (%)*	Average concentration in soil (%)*	Soil:crust ratio
Oxygen	46.5	49	1.05
Silicon	27.8	32	1.16
Aluminium	8.1	7.1	0.88
Iron	5.1	3.8	0.75
Calcium	3.6	1.4	0.39
Sodium	2.8	0.6	0.21
Potassium	2.6	0.8	0.31
Magnesium	2.1	0.5	0.24

* Values taken from W.L. Lindsay (1979) *Soil Chemical Equilibria*, pp. 7–8. John Wiley & Sons, New York.

Rocks are categorised into three major groups: igneous, sedimentary or metamorphic.

Igneous rocks are formed from molten magma within the Earth's crust and have interlocking crystals. *Extruded* igneous rock is formed when the magma appears at the surface and cools quickly, resulting in small grain size (e.g. andesite, basalt, rhyolite). *Intruded* igneous rock cools slowly on its way to the surface and results in large grain size (e.g. gabbro, granite).

Sedimentary rocks are formed from the weathering products of other rocks, or from a biogenic origin, and have non-interlocking crystals. Examples are sandstone (quartz sand), shale and mudstone (clay), limestone (calcium carbonate, of biological origin – shells and skeletons of marine organisms), conglomerate (small stones of various mineralogical make up). In each case the components have been deposited by sedimentation and, over time, compacted and cemented to form rock.

Metamorphic rocks are formed by recrystallisation as a result of high temperature and/or high pressure acting on igneous or sedimentary rocks, usually making them more resistant to weathering. For example, quartzite is formed from sandstone, slate from shale and mudstone, marble from limestone.

Unconsolidated materials are partially weathered rocks that have been transported in various ways and deposited on the Earth's surface. The most important groups of such materials are:

- *glacial deposits*, resulting from the action of ice
- *alluvium*, deposited from water
- *aeolian* or wind-blown deposits.

The rock types from which they originate determine their composition and characteristics.

Primary Silicate Minerals

The *primary minerals* in the surface environment are those that survive weathering of the rocks and unconsolidated material in a relatively unchanged form. They are distinguished from *secondary minerals*, which are formed by various surface chemical processes. Most primary minerals are silicates (Table 1.2), based on the SiO_4^{4-} tetrahedron (Figure 3.1), but in some cases the silicon atom may be replaced by aluminium by the process of *isomorphous substitution* (see Topic 3), causing an imbalance of charge. This is neutralised by cations such as K^+, Na^+, Ca^{2+}, Mg^{2+} and Fe^{2+}.

The silica tetrahedra join together by sharing oxygen atoms to form different silicate structures with varying susceptibility to weathering:

- *Framework silicates*, such as quartz and the feldspars, result from the sharing of all four oxygens by adjacent tetrahedra. The basic unit is $(SiO_2)^0$.
- *Sheet silicates*, such as mica, are formed by the sharing of three oxygens by each silica tetrahedron. The basic unit is $(Si_2O_5)^{2-}$.
- *Chain silicates*, such as amphiboles (double chains) have the tetrahedra sharing two or three oxygens alternately; in pyroxenes (single chains) there is sharing of two oxygens. The basic unit is $(Si_4O_{11})^{6-}$ for double chains and $(SiO_3)^{2-}$ for single chains.
- *Isolated silicates (or orthosilicates)*, such as olivine, consist of individual tetrahedra. The basic unit is $(SiO_4)^{4-}$.

Table 1.2 Structures of common primary silicate minerals.

Quartz	SiO_2
Feldspars	
Microcline	$KAlSi_3O_8$
Albite	$NaAlSi_3O_8$
Anorthite	$CaAlSi_2O_8$
Micas	
Muscovite	$KAl_2(Si_3Al)O_{10}(OH)_2$
Biotite	$K(Mg,Fe)_2(Si_3Al)O_{10}(OH)_2$
Amphiboles	
Hornblende	$(Na,Ca)_2(Mg,Fe,Al)_5(Si,Al)_8O_{22}(OH)_2$
Pyroxenes	
Enstatite	$MgSiO_3$
Augite	$Ca(Mg,Fe,Al)(Si,Al)_2O_6$
Olivine	$(Mg,Fe)_2SiO_4$

Silicates lower down the list in Table 1.2 are more readily weathered, as fewer silicon–oxygen bonds need to be broken to release silicate. Quartz and feldspars especially, and also muscovite mica in temperate regions, are common minerals in the coarse particle size fractions (0.002–2 mm) of many soils and sediments (see Topic 16). Biotite mica, amphiboles, pyroxenes and olivine are much more easily weathered. These minerals contain high amounts of Fe and Mg compared to quartz, feldspars and muscovite mica, and are known collectively as the *ferromagnesian minerals*.

Non-Silicate Minerals

Examples of important non-silicate minerals are shown in Table 1.3.

Secondary Minerals

Secondary minerals are formed by the alteration of a primary mineral or by reaction between soluble components released from a primary mineral reacting together to form a new mineral. The main types of secondary mineral are aluminosilicate clay minerals, hydrous oxides and short-range order (amorphous) aluminosilicates. The first two groups of minerals are discussed separately in Topics 3 and 4.

The short-range order (amorphous) aluminosilicates are the least well understood of the secondary minerals. They have small particle size with no regular crystal structure and a very high surface area, making them highly reactive. They have both a fixed negative charge, arising from isomorphous substitution in Si tetrahedral sheets, and a variable, pH-dependent charge similar to that on clay edges and hydrous oxide surfaces. The two main minerals in this group, allophane and imogolite, are found in young soils and soils of volcanic origin.

Table 1.3 Structures of some common non-silicate minerals.

Carbonates	
Calcite	$CaCO_3$
Dolomite	$MgCO_3 \cdot CaCO_3$
Sulfides	
Pyrite	FeS_2
Galena	PbS
Sphalerite	ZnS
Oxides	
Silica	SiO_2
Hematite	Fe_2O_3
Rutile	TiO_2
Phosphates	
Apatite	$Ca_5(PO_4)_3(OH)$
Sulfates	
Gypsum	$CaSO_4$
Barytes	$BaSO_4$
Halides	
Halite	$NaCl$
Fluorite	CaF_2

2. Weathering Processes

Processes that act on rocks and minerals bringing about changes in their composition and characteristics are collectively known as weathering. These can be physical, chemical or biological processes.

Physical Weathering

Initially, physical weathering dominates, causing fragmentation of rock, a decrease in particle size and increase in surface area, but no chemical change. It is caused by a number of actions:

- *Abrasion* of the rock due to the action of ice, water and wind, leading to physical disintegration
- *Temperature change* due to diurnal temperature variation and freeze–thaw cycles, creating significant pressures that cause rock to split along lines of weakness
- *Crystallisation* of some salts, causing swelling in cracks and fissures, which exerts pressure on the rock
- Once plants become established, *roots* tend to grow into existing fissures and channels and can exert significant pressure

The key point about physical weathering is that it increases the surface area of rock that is available for attack by various chemical and biological weathering processes.

Transport of Partially Weathered Rock

Partially weathered rock that has been broken down by physical processes can be further weathered by chemical processes *in situ*, or can be transported by various means and deposited elsewhere. The main agencies by which transport occurs are shown in Table 2.1.

Chemical Weathering

This is the reaction between natural waters and the minerals in rocks causing partial or complete dissolution of the rock and formation of a new mineral. A number of types of reaction can occur.

Dissolution – Soluble Minerals Dissolve in Water

Minerals such as gypsum ($CaSO_4.2H_2O$) are highly soluble and persist only in arid and semi-arid regions, where water is limiting. In temperate and tropical regions there is sufficient water to cause dissolution and loss of gypsum:

$$CaSO_4.2H_2O \rightleftharpoons Ca^{2+} + SO_4^{2-} + 2H_2O$$

Quartz is a much less soluble mineral than gypsum, but even this is slightly soluble:

$$SiO_2 + 2H_2O \rightleftharpoons H_4SiO_4$$

This reaction is driven by the loss of the soluble silicic acid. In temperate regions there is insufficient water and, as weathering has proceeded over a relatively short time since the last ice age, there has been no significant loss of silica, so quartz is a dominant mineral in the coarse fraction of many soils. In the humid tropics, however, the much larger volume of water passing through the soil over a much longer timescale and the higher temperature have resulted in significant, if not complete, loss of silica.

Table 2.1 Transport processes for weathered rock.

Ice	
Glacial drift	Moved and deposited by glaciers e.g. drumlins
Till	Produced by grinding action of glaciers, contains angular stones
Fluvio-glacial deposit	Deposited from melting ice water, some degree of water sorting and rounded stones
Water	
Alluvium	Deposited from flowing water in streams and rivers, sorted by size and density, contains rounded stones
Lacustrine deposits	Sediment deposited from lake water
Beach deposits	Deposited by wave action, usually sands
Wind	*Aeolian* deposits, e.g. desert sands, loess
Gravity	
Colluvium	Material deposited at the foot of slopes due to processes of *solifluction* – wet soil sliding over frozen ground or *creep* – slow movement of soil downslope

Topic 2 Weathering processes

Differential leaching of soluble ions from rock can result in formation of a new mineral. For example, loss of potassium ions, hydroxyl ions and silicic acid from feldspar alter it to the clay mineral kaolinite:

$$4KAlSi_3O_8 + 22H_2O \rightleftharpoons Al_4Si_4O_{10}(OH)_8 + 8H_4SiO_4 + 4K^+ + 4OH^-$$
$$\text{Feldspar} \qquad\qquad\qquad \text{Kaolinite}$$

Acid Hydrolysis – Carbonic Acid Attacks the Minerals

The water that acts upon rocks and minerals is in equilibrium with CO_2 in the atmosphere, which dissolves to form carbonic acid. Unpolluted rainwater has a pH of approximately 5.7 (see Topic 11), whereas water in soil pores may be exposed to air containing a higher partial pressure of CO_2 than the free atmosphere, and hence soil water may be more acidic. The attack on minerals by this weak carbonic acid is often the major chemical weathering process. For example, acid hydrolysis of calcium carbonate yields calcium and bicarbonate ions:

$$CaCO_3 + H_2CO_3 \rightleftharpoons Ca^{2+} + 2HCO_3^-$$

Acid hydrolysis of the primary mineral microcline feldspar results in release of some soluble components (silicic acid, potassium ions and bicarbonate ions) and alteration of the solid phase to kaolinite (compare this with the dissolution reaction above):

$$4KAlSi_3O_8 + 22H_2O + 4CO_2 \rightleftharpoons Al_4Si_4O_{10}(OH)_8 + 8H_4SiO_4 + 4K^+ + 4HCO_3^-$$
$$\text{Feldspar} \qquad\qquad\qquad\qquad \text{Kaolinite}$$

Oxidation – Oxygen Attacks the Minerals

For those elements that can exist in more than one valence state, oxidation may be a major reaction in the chemical weathering process (see Topic 12). Iron and manganese are the most important elements that behave in this way. For example, iron in the ferromagnesian minerals is in the Fe(II) state, which is oxidised to Fe(III) when released from the mineral. This can cause changes in the charge balance, requiring other ions to be lost. Formation of iron oxide can cause physical disruption to the mineral.

Minerals formed under anaerobic conditions, such as pyrite (FeS_2) will oxidise when exposed to air:

$$2FeS_2 + 7O_2 + 2H_2O \rightleftharpoons 2Fe^{2+} + 4SO_4^{2-} + 4H^+$$
$$4Fe^{2+} + O_2 + 4H^+ \rightleftharpoons 4Fe^{3+} + 2H_2O$$

Biological Weathering

Once biological activity starts, a new set of important weathering reactions can occur due to the release of organic compounds by colonising organisms and plant roots. These compounds are often organic acids that can affect the pH of their environment and form chelates with cations in the minerals.

Chelation – Formation of a Stable Complex Between a Metal Ion and an Organic Molecule

$$R\begin{matrix}COOH\\ \\COOH\end{matrix} + M^{x+} \rightleftharpoons R\begin{matrix}COO\\ \\COO\end{matrix}M^{(x-2)+} + 2H^+$$

The process of chelation is discussed in Topic 10. The organic acids released by lichens, algae, mosses and plant roots cause loss of metal ions from minerals due to the increased solubility of many metal ions at low pH (see Topic 9). These ions can then be chelated by the organic molecule, which often acts to maintain them in solution and therefore allows them to be washed away from the weathering rock.

Consequences of Weathering and Weatherability of Minerals

Because minerals have varying weatherability (i.e. susceptibility to weathering) the types of minerals currently found in the surface environment reflect the nature of their specific environment. So readily weathered minerals, such as halides and sulfates, persist only in arid environments. In highly weathered tropical environments the high temperatures and rainfall have resulted in loss of all but the most resistant minerals, such as iron oxides and kaolinite; most silicates have been lost by the slow leaching of silicic acid. In temperate environments silicates tend to persist and are the main minerals in most surface environments.

3. Clay Minerals

In temperate regions, the clay minerals are the most important secondary minerals. They are sheet aluminosilicates and their properties dominate most soils and sediments in these regions.

All clay minerals are made up of silica tetrahedra (Figure 3.1a), which form *tetrahedral sheets* by the sharing of oxygens between adjacent silicons, and alumina octahedra (Figure 3.1b), which form *octahedral sheets* by the sharing of oxygens or hydroxyls between adjacent aluminiums.

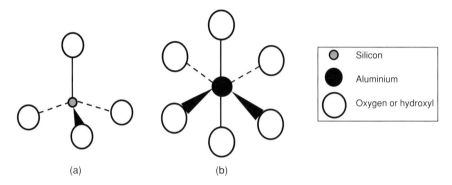

Figure 3.1 Building blocks of clay minerals: (a) silica tetrahedron, (b) alumina octahedron.

Clay *unit layers*, of which there are two basic types, are formed by the sharing of oxygens between tetrahedral and octahedral sheets:

- *1:1 clay* – one tetrahedral silica sheet and one octahedral alumina sheet
- *2:1 clay* – two tetrahedral silica sheets sandwiching one octahedral alumina sheet.

Unit layers are held together in various ways to produce a clay crystal. These are regular, rigid systems and the distance between equivalent points in adjacent unit layers, the *c spacing* or *basal spacing*, can be measured by X-ray diffraction and is used to identify the clay minerals (Figure 3.2).

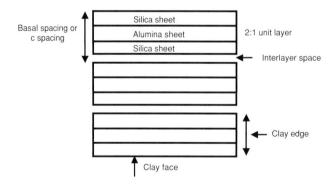

Figure 3.2 Structure of a 2:1 clay mineral (a 1:1 mineral has only one silica sheet and one alumina sheet).

Within the sheets, during the formation of the clay, Si or Al may be replaced by a different element by the process of *isomorphous substitution*. Common replacements in the clay minerals are Al^{3+} for Si^{4+} in the tetrahedral sheet and Mg^{2+} or Fe^{2+} for Al^{3+} in the octahedral sheet. Substitution of an element by one of a lower valency causes *a permanent negative charge* on the surface of the clay mineral. The much smaller (in terms of area) clay edges have a *pH dependent variable charge* similar to that found on the hydrous oxides (see Topic 4). The overall charge on the clays is negative and is neutralised by cations attracted to the surface. It is termed the *cation exchange capacity* (CEC) of the clay, and is expressed in units of centimoles of monovalent charge per kilogram ($cmol_c$/kg).

Common Clay Minerals

Kaolinite is the commonest 1:1 clay mineral. A small amount of isomorphous substitution of Al for Si in the tetrahedral sheet results in a charge of ≤ 0.005 mol negative charge per unit cell. The unit layers are held together by hydrogen bonds between oxygen atoms on the tetrahedral face and hydroxyls on the adjacent octahedral face, resulting in a low CEC (3–20 $cmol_c$/kg) and low surface area (5–100 m^2/g) and a fixed c spacing of 0.7 nm. Unit layers stack up to form hexagonal crystals which are typically 0.05–2 μm thick, so the variable, pH-dependent charge on the clay edge is relatively more important in kaolinite than the other clays (see Figure 3.3).

Illite (a hydrous mica) is a 2:1 clay mineral with isomorphous substitution, mainly in the tetrahedral sheet (Al^{3+} for Si^{4+}), resulting in a charge of 1.5 mol negative charge per unit cell. Most of this charge is neutralised by K^+ ions in the interlayer space. The distribution of oxygen atoms on the tetrahedral face allows K^+ ions to sit very close to the clay surface, which results in a very strong bond between unit layers. The consequence of this is a low CEC (10–40 $cmol_c$/kg), low surface area (100–200 m^2/g) and a fixed c spacing of 1.0 nm (see Figure 3.4).

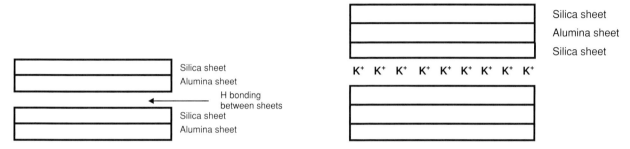

Figure 3.3 Structure of kaolinite.

Figure 3.4 Structure of illite.

Vermiculite, a 2:1 clay mineral, is also a hydrous mica, with isomorphous substitution in the tetrahedral sheet resulting in a charge of between 1.2 and 1.9 mol negative charge per unit cell. In this case the interlayer cations are mainly Mg^{2+} with some Ca^{2+}, more strongly hydrated cations than K^+, so the water of hydration widens the c spacing to up to 1.4 nm, and allows some degree of expansion, resulting in a higher CEC (80–150 $cmol_c$/kg) and surface area (300–500 m^2/g).

Smectites are a group of 2:1 clay minerals with a low degree of isomorphous substitution and low layer charge. Montmorillonite is the commonest smectite in which substitution occurs in the octahedral layer, giving 0.7 mol negative charge per unit cell. Interlayer bonding is weak as the charge originates further from the clay surface, allowing expansion of the clay lattice and easy entry of cations and water molecules into the interlayer space. The CEC (80–120 $cmol_c$/kg) and surface area (700–800 m^2/g) are both high and the c spacing is variable, up to 1.9 nm when Ca^{2+} or Mg^{2+} are the dominant interlayer cations. If sodium is the dominant cation, the unit layers of smectites can be fully dispersed in suspension.

Chlorite is a 2:1:1 clay mineral, with the usual 2:1 unit layer of two tetrahedral sheets and one octahedral sheet held together by a sheet of brucite, a magnesium hydroxide in which about two-thirds of the Mg^{2+} ions are substituted by Al^{3+}, resulting in a positive charge. This neutralises much of the negative charge that arises due to isomorphous substitution in the tetrahedral sheets (2 mol negative charge per unit cell) and forms a strong electrostatic bond between unit layers. There is also hydrogen bonding between hydroxyls on the brucite sheet and oxygens on adjacent tetrahedral sheets. As a result of this strong interlayer bonding, the CEC (10–40 $cmol_c$/kg) and surface area (300–500 m^2/g) are low and the c spacing is fixed at 1.4 nm (see Figure 3.5).

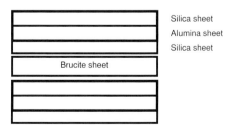

Figure 3.5 Structure of chlorite.

In reality, clay minerals are often not so well defined. It is common for interstratification to occur; for example, intergrades of vermiculite and illite, or smectite and illite, are found (some layers in the clay mineral are illitic in nature, while others more closely resemble vermiculite or smectite). At the edge of an illite crystal some interlayer K^+ ions may have been lost and replaced by other cations, causing a widening of the interlayer space to form a so-called wedge site (see Topic 35).

4. Hydrous Oxides

The term 'hydrous oxides' includes the oxides, hydroxides and oxyhydroxides of aluminium, iron and manganese. These are secondary minerals that form when Al, Fe and Mn are released from primary minerals by weathering (see Topics 1 and 2). They exist in soils and sediments mainly as small particles in the clay sized fraction (<2 μm) (see Topic 16), as coatings on other minerals or as components of larger aggregates (see Topic 17).

Aluminium Oxides

Gibbsite (γ-Al(OH)$_3$) is the most common aluminium oxide. It is grey-white in colour and so tends not to be visible because other, darker coloured materials mask it. It is a relatively persistent mineral and is a major component in highly weathered tropical soils (oxisols). Unlike Fe and Mn, aluminium does not undergo redox changes, existing only as Al(III), so anaerobic conditions do not alter the Al oxides.

Iron Oxides

A number of iron oxides exist, the commonest being *ferrihydrite* (Fe$_2$O$_3$.nH$_2$O), *goethite* (FeOOH) and *hematite* (α-Fe$_2$O$_3$). Iron released from weathering of the ferromagnesian minerals precipitates out of solution either as ferrihydrite or goethite depending on the prevailing environmental conditions (low organic matter and a high rate of Fe release favour ferrihydrite formation). Ferrihydrite is a poorly-ordered oxide of small particle size, whereas goethite is a well-structured mineral and is the commonest of the iron oxides, especially in temperate regions. Hematite forms from ferrihydrite by structural changes to the crystal form, and is commonly found in tropical soils, giving them their bright red colour. Prolonged anaerobic conditions, for example waterlogged soils and sediments, can cause reduction of Fe(III) to Fe(II), which is highly soluble and released from the oxide crystal. Reoxidation initially brings about precipitation of *lepidocrocite* (γ-FeOOH), which slowly converts to goethite if oxidising conditions persist. The orange colour of Fe(III) and green of Fe(II) are seen in gley soils, giving them their characteristic mottled appearance (see Topic 15).

Manganese Oxides

Manganese oxides occur in a large number of forms, often with variable valency, for example *birnessite* [(Na,Ca)(Mn^{3+}, Mn^{4+})$_7$O$_{14}$·2.8H$_2$O], and in mixed oxides, especially with iron. They are the least well understood of this group of minerals. Mn(IV) is readily reduced to Mn(II) under moderately anaerobic conditions, so small concretions of Mn oxide 1–2 mm in diameter are commonly seen in soils and sediments that undergo changes in their redox potential.

Variable pH Dependent Charge

The hydrous oxides have a variable charge, which depends on the pH of their environment (Figure 4.1). This charge develops because of the broken bonds, with unsatisfied electrons, at the oxide surface. Hydrogen ions are taken up or released by the surface and cause the variation in charge. Generally, the charge is positive under more acid (low pH) conditions and negative at high pH. There is one specific pH value at which the overall charge on the oxide surface is zero – *the point of zero net charge*, or *pznc*. When the pH is below the pznc, the net charge on the surface is positive, and conversely above the pznc it is negative. The value of the pznc varies depending on the type of oxide, but broadly Mn oxides have a pznc of about pH 2–4, Fe oxides in the range pH 6–8 and Al oxides in the range pH 7.5–9.5.

Environmental Roles of Hydrous Oxides

The variable charge on the hydrous oxides allows them to interact with other environmental components. This is particularly the case when the oxides are at pH values below their pznc, and so have a net positive charge, when they can bind with

Topic 4 Hydrous oxides

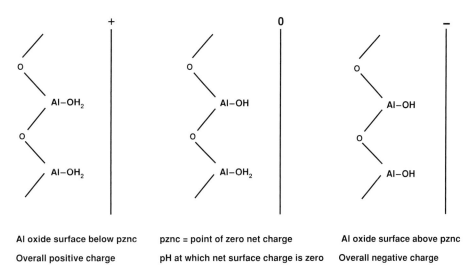

Figure 4.1 Variable, pH dependent charge on an aluminium oxide surface.

negatively charged clay (Topic 3) and humus particles (Topic 5). Such binding is important in creating soil structure and in flocculating solid particles in natural waters, leading to sedimentation (Topic 6).

Hydrous oxides can also attract and bind ions to their surface by the process of ligand exchange (Topic 8). They are highly effective at holding anions such as phosphate, and heavy metals such as zinc, lead and copper. When the net charge is positive, important nutrient ions such as nitrate and sulfate may be held by ion exchange.

5. Humified Organic Matter

Organic Matter Inputs and Humification

Organic matter inputs to the surface environment are predominantly from plant litter, with the amount added per year depending very much on the dominant vegetation type (Table 5.1). In addition, inputs from animal and microbial excretion and decomposition must be considered. The organic materials in these inputs are subject to varying rates of degradation. Simple compounds, such as sugars and amino acids, decompose rapidly as a result of microbial and soil faunal action, whereas complex material, such as lignin and hemicellulose, persists in an unaltered or partially degraded form. Much of the carbon added (about 70%) is lost as CO_2 produced by microbial respiration. The rest is incorporated into the microbial biomass and humified organic matter, which degrades only very slowly.

Humus formation (humification) occurs in two main ways. The more easily decomposable material releases low molecular weight compounds, especially phenolic and amino compounds, which react and polymerise to form humic material. Alternatively, the more resistant organic material, such as lignin, undergoes only partial degradation, and this partially degraded material can react with the low molecular weight compounds released into the soil, again leading to the formation of humus. It is likely that both of these processes operate in the environment.

Table 5.1 Typical annual inputs of carbon to the surface environment under different vegetation types.

Dominant land use	Carbon added [t/ha/year]
Alpine and arctic forest	0.1–0.4
Arable agriculture	1–2
Coniferous forest	1.5–3
Deciduous forest	1.5–4
Temperate grassland	2–4
Tropical rain forest	4–10

Adapted from R.E. White (1997) *Principles and Practice of Soil Science*, p. 37. Blackwell Science, Oxford.

Characteristics of Humified Organic Matter

Humified organic matter, humus and humic substances are all terms used to refer to the resistant organic material that persists in the surface environment. Estimating the age of humic material is possible using the ^{14}C dating method (see Topics 47 and 48). This is based on the fact that a small amount of the carbon fixed by plants in the process of photosynthesis is cosmogenic ^{14}C and that this transfers to the humic material from the plant tissue. ^{14}C undergoes radioactive decay with a half-life ($t_{1/2}$) of 5730 years, so it has proven suitable for measurement of ages up to 50 000 years. There is an assumption, which is reasonably valid, that the rate of production of cosmogenic ^{14}C and the rate of transfer of carbon between different components of the environment have not significantly changed over this time. Using this technique, estimates of mean residence times for carbon in soils of a few hundred to a few thousand years have been made. This is a mean value for all the carbon in soil. Much of the carbon entering soil is rapidly lost as CO_2 due to microbial metabolism, but that which persists can remain for long periods of time, making soil organic matter a major pool of carbon in the environment. One model, which divided the carbon into different forms based on stability of the organic component, recognised five fractions with different turnover times: *decomposable plant material* (<1 year); *resistant plant material* (1–5 years); *microbial biomass* (1–5 years); *physically stabilised organic matter* (50–100 years); and *chemically stabilised organic matter* (2000–3000 years).

The detailed chemical make-up of humified organic matter from different environments may be quite diverse, depending on the dominant type of plant tissue input, but the overall characteristics are similar. Humus comprises dark coloured, amorphous, colloidal polymers built up mainly from aromatic units, with a wide range of molecular weights (100s to 100 000s Da). Figure 5.1 shows the random coiled nature of humic material.

The overall properties are due to the preponderance of carboxylic and phenolic functional groups on the humic polymer. Both of these groups can ionise by losing a hydrogen ion to give a negatively charged group:

(a) $R-COOH \rightleftharpoons R-COO^- + H^+$

(b) $C_6H_5-OH \rightleftharpoons C_6H_5-O^- + H^+$

This is a variable charge, the magnitude of which varies with pH, but is always negative and forms part of the cation exchange capacity of a soil. The pK of carboxylic groups (equation a) is in the range 3–5, and of phenolic groups (equation b) in the range 7–8: this is the pH value at which the group is 50% ionised. Thus carboxylic groups are important in acid soils, while phenolic groups become important above pH 7.

Topic 5 Humified organic matter

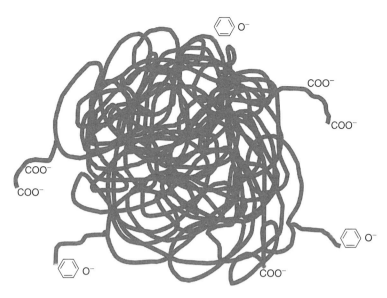

Figure 5.1 Diagrammatic representation of humic material (not to scale).

Fractionation of Humified Organic Matter

Humified organic matter has been traditionally fractionated into a number of subgroups (Figure 5.2). *Humin* and *humic acid* are high molecular weight fractions (>10 000 Da), while *fulvic acid* is the low molecular weight fraction (<10 000 Da). There are some chemical differences between humic acid and fulvic acid, the latter being more acidic and having a higher oxygen content and lower carbon content than humic acid. This reflects the greater number of carboxylic and phenolic groups in the fulvic acid fraction. It is speculated that humic acid forms the less reactive backbone or core of the humic material, while the more reactive fulvic acid is present as side chains branching from the core.

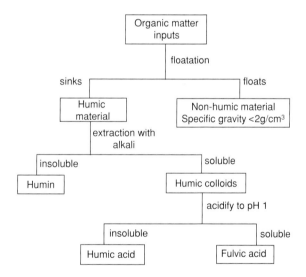

Figure 5.2 Fractionation of humified organic matter.

Chemical Composition of Humified Organic Matter

Although there is considerable variation, the ratio of the major elements C:N:P:S in humic material is approximately 100:10:2:1, forming major reservoirs of N, P and S in the environment, which are released as humified organic matter broken down by soil micoorganisms (see Topics 33, 34 and 35). Identification of N-, P- and S-containing compounds is difficult, and those that are identified tend to be metabolic products, such as nucleotides and vitamins, released into the soil following the death of cells. Broadly, nitrogen is an integral part of the humic molecule, and is released as NH_4^+ ions when the humic material is degraded. Phosphorus and sulfur are more commonly found as P and S esters, which can be released as orthophosphate and sulfate ions by the action of phosphatase and sulfatase enzymes, respectively.

6. Flocculation and Dispersion of Clays, Oxides and Organic Matter

The charged components discussed in Topics 3 (clays), 4 (oxides) and 5 (humified organic matter) can interact with each other and with charged ions and molecules in the environment. Such interactions result in larger units that can sediment out of solution in natural waters, and which in soils help to create a structure.

Flocculation and Dispersion of Clays

Multivalent cations, such as Ca^{2+}, can decrease the repulsion between two negatively charged clay surfaces, allowing the clay unit layers to approach each other sufficiently closely to allow attractive van der Waals' forces to produce a reasonably stable entity – a clay domain (Figure 6.1) – this is called *flocculation* of the clay.

Flocculation depends on the concentration and size of the cation, which determines how closely the layers can approach each other, and the charges of the cation and clay surface. The surface charge depends on the extent and type (octahedral or tetrahedral) of isomorphous substitution. If the cation is present at low concentration, or has a small charge, more water molecules are present between the clay layers, creating a repulsive osmotic pressure, which tends to push the clay layers apart, causing *dispersion*. This can be visualised in Figure 6.2.

The ability of cations to flocculate clay can be ordered on the basis of ionic charge and the size of the *hydrated radius* of the ion (i.e. its effective radius when dissolved in water and surrounded by water molecules – Table 6.1). This order is called the *Hofmeister series*:

$$Li^+ > Na^+ > K^+ \sim NH_4^+ > Rb^+ \sim Cs^+ > Mg^{2+} > Ca^{2+} \sim Sr^{2+} > M^{3+} > M^{4+}$$

Ionic charge and, within a group of ions of the same charge, hydrated ionic radius both increase from left to right. Therefore ions further to the right of the Hofmeister series are more effective at flocculating clay. Concentration is also a factor, and

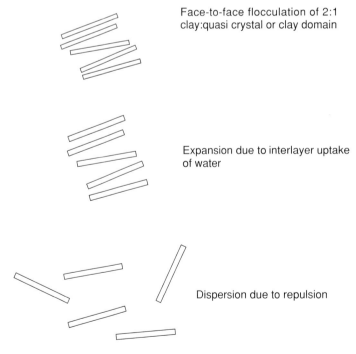

Figure 6.1 Flocculation and dispersion of a 2:1 clay mineral (▭ represents a 2:1 clay unit layer).

Topic 6 Flocculation and dispersion of clays, oxides and organic matter

Table 6.1 Crystal radii and hydrated radii of common mono- and divalent cations.

Ion	Crystal radius (nm)	Hydrated radius (nm)
Li^+	0.068	1.00
Na^+	0.098	0.79
K^+	0.133	0.53
NH_4^+	0.143	0.54
Rb^+	0.149	0.51
Cs^+	0.165	0.51
Mg^{2+}	0.089	1.08
Ca^{2+}	0.117	0.96
Sr^{2+}	0.134	0.96

Data adapted from M.A. Tabatabai and D.L. Sparks (2005) *Chemical Soil Processes*, p. 355. Soil Science Society of America, Madison, WI.

Figure 6.2 Flocculation of clay by cations of different valencies.

approximate flocculation values are: M^+, 25–150 mM; M^{2+}, 0.5–2.0 mM; M^{3+}, 0.01–0.1 mM. In soils, Ca^{2+} and Mg^{2+} are common flocculating ions, whereas Na^+ tends to be a dispersing ion at the concentrations normally found in soil solution. In estuaries and soil flooded by seawater, the high concentration of Na^+ flocculates the clay.

Clay minerals can also flocculate by attraction of the permanently negatively charged surface to the positively charged edge below the pznc (see Topic 4). Clay layers align themselves edge-to-face (Figure 6.3).

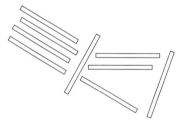

Figure 6.3 Edge-to-face flocculation of clays (▭ represents a 1:1 or 2:1 clay unit layer).

Flocculation Involving Hydrous Oxides and Humified Organic Matter

In many soils, sediments and natural waters, hydrous oxides are below their point of zero net charge (pznc, see Topic 4), and so are positively charged. They are attracted to the negatively charged clay surfaces, often existing as coatings on clays, which can mask the negative sites, i.e. decrease the cation exchange capacity (CEC, see Topic 7). Oxides can also bond with the negatively charged humified organic matter (see Topic 5).

Multivalent cations, such as Ca^{2+}, can act as a bridge between negative sites on different components, e.g. organic matter–clay, organic matter–organic matter.

Significance of Flocculation in the Environment

In soils and sediments, the flocculation processes described above help to stabilise the soil or sediment by building small particles into larger units, creating a structure and a pore system. In natural waters, such as lakes and rivers, interactions of this type bring about sedimentation of solid particles from the water. For example, flocculation rapidly occurs when river sediments enter the much higher ionic strength marine water.

7. Ion Exchange

Definition of Ion Exchange

Ion exchange occurs when ions held at a charged surface by coulombic bonding are replaced by other ions of the same charge present in solution in contact with the surface. This is an important process controlling the behaviour of certain ions in soil/sediment/water systems.

In Figure 7.1, A^+ and B^{2+} ions in solution replace X^+ and Y^{2+} ions held on a negatively charged solid phase surface. Cation exchange is the major reaction that occurs in the natural environment. Anion exchange, negatively charged ions held at positively charged surfaces, can occur, but is of minor importance.

Ion exchange occurs between the surface of soil particles and soil water or between the surfaces of suspended sediment and river or lake water.

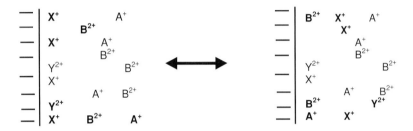

Figure 7.1 Cation exchange at a clay surface (ions in bold are undergoing ion exchange).

Solid Phase Surfaces onto Which Cation Exchange Can Occur

There are three types of surface on to which cation exchange can occur.

Aluminosilicate clays (see Topic 3) – Aluminosilicate clays are commonly found in most soils and sediments, and as a result of erosion can be washed out into natural waters. They have a permanent negative charge on the face of the clay lattice due to isomorphous substitution within the lattice, and a pH variable charge at the lattice edge due to broken bonds.

Humified organic matter (see Topic 5) – Humified organic matter is the dark, amorphous organic material found in soils, sediments and waters. It is formed by microbial breakdown of dead tissue and the synthesis of humified organic matter from the degradation products. The presence of certain functional groups on the humic polymer give rise to both positive and negative charges, but the overall charge is negative due to the preponderance of carboxylate and phenolate groups. The magnitude of this charge varies with pH.

Hydrous oxides (see Topic 4) – The important surfaces are those of the hydrous oxides of Al, Fe and Mn, which are common secondary minerals in many soils and sediments, and which can be found suspended in natural waters. The charge on these surfaces is pH dependent, being positive at low pH and negative at high pH.

The overall negative charge from these surfaces is the *cation exchange capacity* (CEC) of the soil or sediment: that is its ability to hold cations at negative sites by coulombic attraction (see Topic 3).

Factors Affecting the Ion Exchange Process

The order of ions given in the Hofmeister series in Topic 6 for flocculating ability also holds for ion exchange preference. For ion exchange in general:

(1) More highly charged ions are held at exchange surfaces in preference to lesser charged ions

$$M^{3+} > M^{2+} > M^+$$

(2) For ions of the same charge, smaller ions are held in preference to larger ions

$$Cs^+ > Rb^+ > K^+ > Na^+ > Li^+$$

Increasing hydrated radius $\longrightarrow$

(3) Ionic concentration must also be taken into account. In highly concentrated systems, e.g. seawater, the presence of a large amount of an ion can override the above two factors.

Ions Involved in Ion Exchange

Although in theory any cation could be held at a negative surface, in practice ion exchange is important only for ions that are found in environmental systems in relatively high concentrations. In soils, for example, the dominant cations are Ca^{2+}, Mg^{2+}, K^+ and Na^+ (collectively known as the exchangeable cations or exchangeable bases); in managed agricultural soils NH_4^+ may be important. The exchange sites on the clays, organic matter and oxides are filled mainly with these cations. Trace elements, such as Zn^{2+} and Cu^{2+}, are present in much smaller concentrations and so tend to be displaced by the macro cations. In a typical agricultural soil of approximate pH 6.5, the distribution of exchangeable cations on the exchange sites would be roughly 80% Ca^{2+}, 15% Mg^{2+} and 5% K^+ and Na^+. Leaching will cause removal of the exchangeable ions and, if they are not replaced by weathering of the soil minerals, the soil will acidify. Once the soil pH falls below about 5, ions such as Mn^{2+} and Al^{3+} are held on exchange sites.

The *% base saturation* is a measure of a soil's ability to supply macronutrient cations such as Ca^{2+}, Mg^{2+} and K^+ to plants:

$$\% \text{ base saturation} = \frac{\sum (\text{exchangeable } Ca^{2+}, Mg^{2+}, K^+ \text{ and } Na^+) \times 100}{\text{CEC}}$$

(Exchangeable cations and CEC are measured in units of $cmol_c/kg$.) Thus a well managed agricultural soil would have a % base saturation of >80%, while in an acidic upland soil the value could be <20%.

Fixation of Ions

In some types of clay mineral, exchangeable cations may be trapped by being 'fixed' between the unit layers of the clay. This is especially important for K^+ in hydrous micas such as illite (see Topics 3 and 35). Ions fixed in this way are not freely exchangeable with ions in solution.

Examples of the Role of Ion Exchange in Environmental Processes

Leaching of Nitrate versus Ammonium

Two chemical species of nitrogen are commonly found in the surface environment: nitrate (NO_3^-) and ammonium (NH_4^+). Ammonium is released by the breakdown of humified organic matter (see Topic 5), and is added to soils as fertiliser (as is nitrate). In aerobic environments of pH > 5.5, ammonium is microbially transformed to nitrate; thus in an agricultural system, nitrate will be the dominant form. Cation exchange can hold ammonium ions in a soil, but there is no such control for nitrate, which is readily leached from the soil into ground- and surface-waters. Nitrate leaching is a serious environmental problem in areas of intensive agriculture (see Topic 39).

Chernobyl Radiocaesium

Following the explosion at the Chernobyl nuclear reactor in April 1986, large quantities of $^{137}Cs^+$ ions were introduced into soils in many European countries. Caesium ions are strongly held by clay minerals by ion exchange and can also be fixed between the unit layers of some clays. It was therefore expected that the radiocaesium would be immobilised when it entered soil. This was the case where it was deposited in mineral soils where ion exchange and fixation by clay minerals would dominate. Where the radiocaesium fell on organic soils with low clay contents, it was less strongly held by the functional groups in the organic matter, and it could not be fixed. This caesium remained in an easily bioavailable form and readily entered the food chain.

8. Adsorption

Definition of Adsorption

Adsorption is a process which can control the distribution of ions and molecules between the solution phase and the solid phase surface of soil or sediment particles. Bonding to the surface is by:

(1) covalent bonds, in a process called *ligand exchange* or *chemisorption*
(2) *hydrogen bonds*, when a hydrogen atom acts as a bridge between two electronegative atoms in different compounds, or
(3) *van der Waals' forces*, weak, non-specific interactions arising from charge separation in molecules.

Hydrogen bonds and van der Waals' forces are individually weak, but their effect is additive. If interactions occur at a number of points the overall adsorption can be strong. (Contrast with ion exchange, Topic 7, involving coulombic bonds only.)

Ligand Exchange (Chemisorption)

The important ions held by this process are certain anions and some trace metals; the main surfaces involved are the hydrous oxides. The ion being chemisorbed replaces a water molecule or hydroxyl ion (OH^-) from the surface (Figure 8.1).

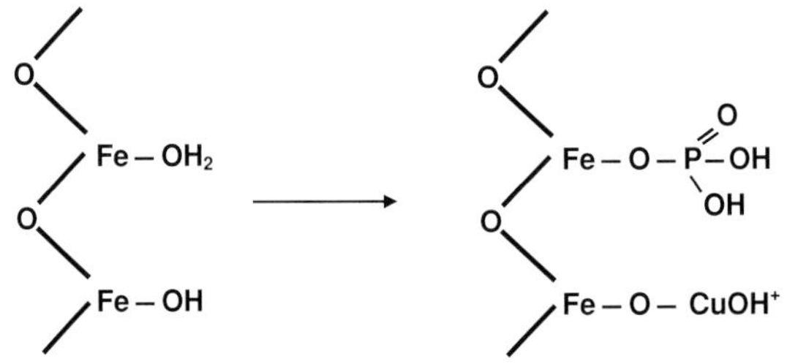

Figure 8.1 Chemisorption of phosphate and copper ions onto an iron oxide surface.

The ions involved in chemisorption are anions of weak acids and hydrolysable cations, both of which can change their chemical speciation by gain or loss of a hydrogen ion (H^+):

$$H_nA \rightleftharpoons H_{n-1}A^- \rightleftharpoons H_{n-2}A^{2-} \quad \text{etc.}$$

$$[M(H_2O)_n]^{x+} \rightleftharpoons [M(H_2O)_{n-1}OH]^{(x-1)+} \rightleftharpoons [M(H_2O)_{n-2}(OH)_2]^{(x-2)+} \quad \text{etc.}$$

The pK values for these reactions (the pH at which there is 50% of each chemical species involved) is important in determining the pattern of chemisorption. For anions, the plot of amount of ion sorbed against pH shows a peak or an inflexion point at the pK value, e.g. for phosphate (see Figure 8.2):

$$H_3PO_4 \rightleftharpoons H_2PO_4^- + H^+ \rightleftharpoons HPO_4^{2-} + H^+ \rightleftharpoons PO_4^{3-} + H^+$$
$$\text{p}K_1\ 2.1 \qquad \text{p}K_2\ 7.2 \qquad \text{p}K_3\ 12.3$$

For cations, the graphs are steep S-shaped curves, with the greatest rate of change in chemisorption taking place 3–4 pH units below the pK value. For example, for copper (see Figure 8.3):

$$[Cu(H_2O)_6]^{2+} \rightleftharpoons [Cu(H_2O)_5(OH)]^+ + H^+$$
$$\text{p}K\ 8.0$$

Adsorption by Hydrogen Bonds and van der Waals' Forces

This is particularly important for the adsorption of organic molecules, such as pesticides and natural humic substances, onto the surfaces of clays, organic matter or hydrous oxides. Hydrogen bonds are formed when a hydrogen atom is shared between two electronegative atoms. For example, in Figure 8.4(a), a hydrogen bond is shown between the nitrogen atom of one humic

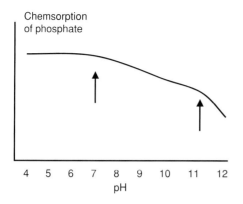

Figure 8.2 Variation with pH of chemisorption of phosphate by a hydrous oxide (arrows show inflection points in the graph, lying at pH values just below pK_2 and pK_3).

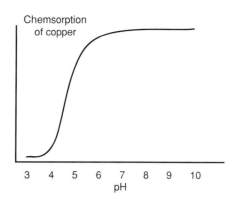

Figure 8.3 Variation with pH of chemisorption of copper by a hydrous oxide.

molecule and the oxygen atom of another; in Figure 8.4(b), the hydrogen bond is between oxygen atoms in a humic molecule and on the surface of an iron hydrous oxide. Van der Waals' forces result from dipole–dipole interactions between two different components due to charge separation within the molecule; for example, between groupings in a humic molecule and a clay surface. In both cases, a large number of interactions result in a strong bond between the two components.

Langmuir Equation

Chemisorption is often described by use of the Langmuir equation:

$$x/x_m = Kc/(1 + Kc)$$

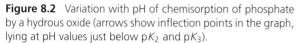

Figure 8.4 Examples of hydrogen bonding.

where c is the concentration of chemisorbed ion in solution at equilibrium, x is the amount of the ion chemisorbed onto the soil, x_m is the maximum amount of ion chemisorbed on to the soil and K is a constant related to bonding. The equation is often used in the linearised form:

$$c/x = c/x_m + 1/Kx_m$$

c can be measured and, knowing the initial concentration and volume of the solution containing the chemisorbed ion, x can be calculated. By plotting c/x against c, a straight line will result (if the Langmuir equation is a suitable model for the process), which will have a slope of $1/x_m$ and an intercept of $1/Kx_m$, allowing the theoretical maximum chemisorption of the ion to be calculated. In practice, application of the Langmuir equation is often not strictly valid, as a straight line relationship is not found, but it is commonly used to obtain a value for x_m.

Examples of the Role of Adsorption in Environmental Processes

Phosphate, silicate, borate, arsenate, selenite, chromate and fluoride are anions for which ligand exchange is important. Nitrate, chloride, bromide and perchlorate are not held, while sulfate and selenate may be weakly held. As a consequence, leaching of nitrate and sulfate from soil in drainage water can be significant, but very little phosphate is lost in solution. Of the trace metals, Co, Cu, Ni and Pb are strongly held on oxide surfaces by chemisorption, but the process is much less important for Cd and Zn.

Hydrogen bonds and van der Waals' forces are particularly important for binding organic molecules, especially onto clay and oxide surfaces. This may be either natural humic material, in which case such interactions help to stabilise soils and sediments, or xenobiotic organics such as pesticides, which can be immobilised and detoxified.

9. Solubility Processes

Definition of Solubility and Precipitation

The concentration in solution of some ions in environmental systems is controlled by the presence of poorly soluble solid phase components. This process can be considered as *solubility* – solid phases in contact with water will eventually come into equilibrium with characteristic concentrations of their constituent ions. This can be represented by the following equation:

$$AB \rightleftharpoons A^+ + B^-$$

where the solid AB dissolves to form ions A^+ and B^- in solution. Alternatively, if the equation is reversed, the process can be considered as *precipitation*: when the concentrations of ions in solution exceed a certain critical value, the ions form a compound which precipitates out of solution as a new solid phase.

The equilibrium constant for the above reaction is

$$K = \frac{[A^+][B^-]}{[AB]}$$

Because the solid phase is poorly soluble, the concentration of AB does not change appreciably and so this can be written as:

$$K_{sp} = [A^+][B^-]$$

where K_{sp} is the solubility product of the solid phase component.

Rates of dissolution can vary greatly (see the examples given in Topic 2 for gypsum and quartz); a soluble mineral such as gypsum will readily dissolve on contact with water, while quartz is so poorly soluble that it persists in many environments and is only lost by dissolution and leaching over a very long timescale.

Precipitation is a relatively much faster process. If the conditions are right (for example, high pH for Al(III) and Fe(III) or redox potential for Fe(II/III) and S(II/VI) in the examples given below), or if the solubility product is exceeded, then a new solid phase forms quickly and the constituent ions are lost from solution.

Examples of the Role of Solubility/Precipitation in Environmental Processes

Good examples of this are Fe(III) and Al(III), both of which are highly insoluble except at very acid pH values, and which are controlled by the solubility of the hydrous oxides; iron(II), in contrast, is much more soluble. Figure 9.1 shows the relative solubilities of Al(III), Fe(III) and Fe(II), assuming precipitation as their hydroxides. Al(III) is insoluble above pH 4–5, when it precipitates as aluminium oxide. Concentrations of Al^{3+} in solution become significant at lower pH and so aluminium toxicity is a problem in acidic soils and water: acid rain, for example, causes release of Al^{3+} ions from some soils and leaching into natural waters (see Topic 52). Fe(III), the oxidised form of iron, is insoluble above pH 2 under aerobic conditions and precipitates as an Fe(III) oxide. Fe(II), the reduced form of iron, is soluble up to about pH 7.5, but only under moderately anaerobic conditions. Deposits of orange iron oxides are often seen in outflows of water from areas of peat because the Fe(II), which is present under the reduced conditions in the peat, is oxidised to Fe(III) when it is washed out into a more aerobic

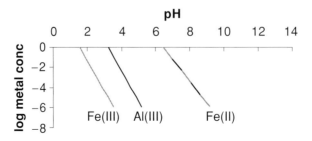

Figure 9.1 Solubility lines for Fe(III), Al(III) and Fe(II).

environment. Similarly, iron leached from colliery spoil or from abandoned coal mines in acid mine drainage, precipitates iron oxide (ochre) when it is washed into higher pH environments such as streams and rivers (see Topic 46). Thus solubility of iron in natural systems is controlled by both redox potential and pH.

Sulfur is another element for which solubility has important effects on its mobility within the environment. The oxidised form, sulfate SO_4^{2-} (S(VI)), is highly soluble and accumulates in soils and sediments only in arid areas where there is insufficient rainwater to wash it out. Under anaerobic conditions, however, sulfate is reduced to sulfide S^{2-} (S(II)), which may exist as hydrogen sulfide gas (H_2S) or, more commonly, as an insoluble metal sulfide precipitate. Iron sulfide (FeS) is often seen as a black layer, streaks or concretions in highly reduced soils and sediments. Thus the mobility of sulfur in the environment is controlled by redox potential. The acid mine drainage mentioned above is a consequence of the oxidation of pyrite (FeS_2), which is formed under the anaerobic conditions from which coal is produced, but is unstable in aerobic environments.

10. Complexation and Chelation by Organic Matter

Definition of Complexation and Chelation

Metal ions in solution are surrounded by a shell of water molecules. Complexes are formed when one or more of these water molecules is replaced by another molecule (called a *ligand*). Organic molecules found in the environment commonly form complexes with metal ions in solution. Pairs of electrons are donated to the metal by electronegative atoms in the organic molecule, forming a coordinate bond. Particularly important are bonds formed by oxygen-, nitrogen- and sulfur-containing groups in humified organic matter. Humic compounds can react with metal cations to form organo-metallic complexes. The reaction of a metal ion (M^{x+}) with a ligand (L) on humified organic matter (OM) can be represented as follows:

$$M^{x+} + OM-LH_y \rightleftharpoons OM-LM^{(x-y)+} + yH^+$$

If more than one bond is formed between the organic molecule and the metal ion, the complex is called a *chelate* and the process *chelation*. This confers much greater stability on the complex than a single bond.

$$R\begin{array}{c}COOH\\ \\COOH\end{array} + M^{x+} \rightleftharpoons R\begin{array}{c}COO\\ \\COO\end{array}M^{x-2} + 2H^+$$

For a reaction between a metal and a ligand

$$M + L \rightleftharpoons ML$$

(where M is the metal ion, L the organic ligand and ML the metal complex) the stability of the complex or chelate formed can be expressed as the equilibrium, or formation, constant K:

$$K = [ML]/[M][L]$$

When more than one ligand bonds to a metal, a series of stepwise reactions can be considered, e.g. for a Cu^{2+} ion binding with two ligands:

$$Cu^{2+} + L \rightleftharpoons CuL^+ \qquad K_1 = [CuL^+]/[Cu^{2+}][L]$$
$$CuL^+ + L \rightleftharpoons CuL_2 \qquad K_2 = [CuL_2]/[CuL^+][L]$$

The overall constant $K_i = [CuL_2]/[Cu^{2+}][L]^2 = K_1 K_2$.

While experimentally [M] can be measured and [ML] calculated, it is difficult to quantify [L], given the polymeric and chemically diverse nature of humified organic matter (see Topic 5).

Ligands and Ions Commonly Involved in Complexation/Chelation

The important functional groups in humified organic matter that contain atoms with lone pairs of electrons that can act as ligands are:

- *Oxygen containing groups*: enolate $-O^-$; carboxylate $-COO^-$; carbonyl $C=O$; ether $-O-$
- *Nitrogen containing groups*: amine $-NH_2$; azo $-N=N-$; ring N
- *Sulfur containing groups*: thiol $-SH$; sulfide $-S-$.

The importance of ions involved in these reactions is often shown as a stability order. Although the exact order can change depending on the type of humic material and the presence of different ligands, a typical order is as follows:

$$Pb > Cu > Ni > Fe \sim Al > Zn > Mn > Cd > Ca$$

Examples of the Role of Complexation/Chelation in Environmental Processes

As can be seen from the order of stability of metal–humic complexes given above, complexation and chelation are important for heavy metals such as Cu^{2+} and Pb^{2+}, but not as important for Zn and Cd. This is borne out in the observation that complexed Cu and Pb compounds are important in soils, sediments and natural waters, whereas other interactions (ion exchange, adsorption) tend to control the mobility or bioavailability of Zn and Cd. Complexation of Fe^{3+} and Al^{3+} is an important process in the formation of certain soils (podzols, see Topic 15).

Complexation by low molecular weight organic compounds, such as those in root exudates, can act to bring metals into solution, making them more mobile and bioavailable; however, reaction with high molecular weight humic material is an immobilising process. So for example, copper, which is an essential micronutrient, is often unavailable to crops growing in organic soils as it is strongly complexed onto the humic material, and Cu deficiency may result.

11. pH and Buffering

Definition of pH

pH is defined as the negative logarithm of the hydrogen ion concentration (or, strictly, activity) in solution: $-\log[H^+]$. It is based on a scale of 0–14, which is derived from the equilibrium constant for the dissociation of water.

$$H_2O \rightleftharpoons H^+ + OH^- \qquad K = 10^{-14}$$

Thus $[H^+][OH^-] = 10^{-14}$. In pure water $[H^+] = [OH^-] = 10^{-7}$, so pH = 7 (as does pOH). If the $[H^+]$ increases, i.e. the system becomes more acidic, the pH falls below 7. Alternatively, if $[H^+]$ decreases, the system becomes more alkaline.

In general, the pH of the natural environment lies within the slightly acidic to slightly alkaline range of pH 5–8, although there are exceptions to this. For example, highly organic soils can have a pH as low as 3, while some salt affected soils may have a pH as high as 10.

Measurement of pH

pH is usually measured using a combination reference and glass electrode. For natural waters, the electrode can be immersed into the sample and the pH measured directly, but for soils and sediments a suspension is used. This makes the measurement more complex, as what is actually being measured is the pH of a solution in equilibrium with negatively charged soil or sediment particles (due mainly to charges on the clay or organic matter particles) (see Figure 11.1): some cations are in solution and some on the soil surface. The pH measured depends on the ratio of these two pools of cations, which is affected by solid:solution ratio in the suspension and the nature of the solution used. As H^+ is one of these cations, and the pH is a measure of $[H^+]$ in solution, both the solid:solution ratio (commonly 1:1, 1:2.5 or 1:5) and the supporting solution (commonly deionised water or a dilute salt solution such as 0.01 M $CaCl_2$) must be stated when reporting a pH value for soil or sediment.

It is important also to remember that pH can change considerably with distance and time as environmental conditions change. In a soil, for example, changes can be due to factors such as: waterlogging, organic matter distribution, carbonate distribution, fertiliser distribution, and microbial or root action.

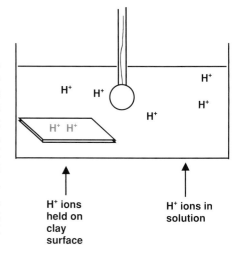

Figure 11.1 Measurement of pH in a soil suspension, with some H^+ ions in solution and some H^+ ions on soil or sediment particles (not to scale).

pH and Buffering

Buffering is the ability of a system to counteract changes in pH when acid or alkali is added. Figure 11.2 shows a typical buffer curve, showing how pH is controlled with addition of acid. H^+ and Al^{3+} ions in solution and on surface exchange sites

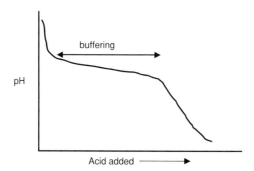

Figure 11.2 Typical buffer curve showing control of pH over a specific pH range.

constitute the *solution and exchangeable acidity* in a soil or sediment, also called the *active acidity*. The *non-exchangeable*, or *reserve acidity* is associated with carboyxl and phenolic groups on humic material (see Topic 5), OH groups on hydrous oxide surfaces and clay edges (see Topics 3 and 4), and polymeric aluminium hydroxides. The balance between solution and exchangeable acidity and non-exchangeable acidity can shift in response to a change in pH:

$$\text{Non-exchangeable acidity} \rightleftharpoons \text{Exchangeable and solution acidity}$$

If soil or sediment pH is increased ($[H^+]$ in solution decreased) the equilibrium shifts to the right and some non-exchangeable acidity is released, whereas if the pH decreases ($[H^+]$ in solution increased) the equilibrium shifts to the left and some acidity becomes non-exchangeable. This represents the *buffer capacity* of a soil or sediment, and prevents large changes in pH.

Natural waters have very little buffering capacity arising from clays, oxides or humic material, as the amounts of these suspended in water are usually low. More important here is buffering due to the presence of certain ions that can take up or release H^+ ions to form different chemical species. Bicarbonate and carbonate are especially important buffers in natural waters. Carbon dioxide dissolves in water to form carbonic acid (H_2CO_3) which ionises to form bicarbonate (HCO_3^-) and carbonate ions (CO_3^{2-}):

$$H_2CO_3 \rightleftharpoons HCO_3^- + H^+ \rightleftharpoons CO_3^{2-} + 2H^+$$

The way in which the different chemical species change is dependent on pH and is shown in Figure 11.3.

Rainwater is buffered at about pH 5.7 as a result of the dissolution of atmospheric CO_2. Seawater is buffered at about pH 8 due to bicarbonate, a highly soluble (and hence mobile) ion in the environment, which accumulates in seawater.

Buffering capacity in water tends to be less than in soils and sediments as the sources of buffering (suspended material or dissolved ions) are finite and often easily overcome. In a well-buffered system, H^+ ions are taken out of solution by the buffer in response to a fall in pH, and conversely released into solution as a result of a rise in pH. The aim of this buffering is to maintain a constant, or reasonably constant, pH environment. Figure 11.4 shows typical buffer curves for (a) soil and (b) seawater. The soil curve (a) shows control of pH over a wide range due to the varying buffer systems (clay, oxides, organic matter), whereas the seawater curve (b) shows little buffering capacity except for a small amount due to bicarbonate ions.

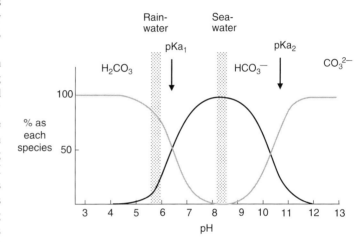

Figure 11.3 Variation in the chemical species of carbonate with pH.

Influence of pH on Environmental Systems

A major influence of pH is its effect on biological activity. Many microorganisms function within a narrow optimal range of pH, and their activity is inhibited under more acidic or alkaline conditions. For example, the nitrifying bacteria *Nitrosomonas* and *Nitrobacter* have a pH optimum in the range 6–8, and are severely inhibited below pH 5.5. Conversely, the iron and sulfur oxidising bacteria of the genus *Thiobacillus* are active only under acid conditions, and are inhibited at pH > 5. In general, bacteria are less tolerant of acid conditions than fungi. Animals too are affected by pH; earthworms, for example, cannot survive below about pH 4.5.

Certain environmental components have a variable, pH-dependant charge. The charge on the humified soil organic matter is negative overall due to dissociation of carboxyl and phenolic groups, but its magnitude varies, being greater at high pH (see Topic 5). The hydrous oxides (Topic 4) and edges of clay minerals (Topic 3) are positively charged at low pH and negatively charged at high pH. In general, therefore, low pH results in a lessening of the negative charge on the surfaces of soil particles. pH has a big influence on the solubility of many important elements in the environment (Topic 9).

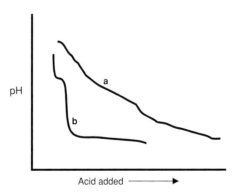

Figure 11.4 Typical buffer curves for (a) soil and (b) seawater.

12. Redox Potential

Definition Of Redox Potential

The redox potential is the measure of the oxidation–reduction state of an environmental system such as a soil, sediment or water body. It is determined by redox reactions, which involve the transfer of electrons from one chemical species to another. Oxidation occurs when electrons are lost; reduction when electrons are gained. (A convenient way to remember this is the mnemonic OIL RIG – Oxidation Is Loss, Reduction Is Gain.) A generalised redox reaction can be written as:

$$\text{Ox} + m\text{H}^+ + n\text{e}^- \longleftrightarrow \text{Red}$$

where Ox and Red are the oxidised and reduced species, respectively. Note also that H^+ ions are involved in the reaction, and so pH affects the redox potential.

Movement of electrons creates an electric potential, determined by the ratio of the activities of oxidised [Ox] and reduced [Red] species, and defined by the *Nernst equation*.

$$E = E^0 - \frac{RT}{nF} \ln \frac{[\text{Red}]}{[\text{Ox}]}$$

where E is the potential in volts, E^0 is the standard electrode potential, R is the universal gas constant 8.314 J/mol/K, T is the absolute temperature in kelvin (K), n is the number of electrons involved and F is the Faraday constant of 96 487 coulombs/mol (the charge when one mole of Ox is reduced)

The Nernst equation is usually simplified by substituting in the constants R and F, assuming a temperature of 25°C (298 K) and converting from natural logarithms to $\log_{10}$, giving

$$E = E^0 + \frac{0.0591}{n} \log_{10} \frac{[\text{Ox}]}{[\text{Red}]}$$

Measurement Of Redox Potential

Redox potential is measured by inserting an inert platinum electrode into the sample; this acquires the electric potential of the sample (Figure 12.1). It is not possible to measure the electric potential of a reaction in isolation; it can be measured only relative to another potential. Formally, redox potential is measured relative to the hydrogen electrode, and so is called E_h. For practical purposes, it is measured against a reference electrode with a known potential, such as the calomel electrode (E_{cal}), which has a potential of 0.248 V at 25°C. This potential has to be added to the measured value:

$$E_h = E_{cal} + 0.248 \text{ V}$$

E_h is measured at the pH of the sample, but is often expressed corrected to pH 7 by subtracting 0.059 V per unit pH up to 7 for samples with a pH below 7 and adding 0.059 V per unit pH down to 7 for samples above pH 7.

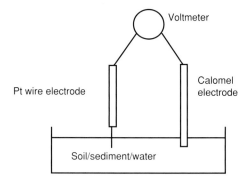

Figure 12.1 Experimental set-up for measurement of redox potential.

Respiration

Respiration provides the energy for the metabolic processes of heterotrophic organisms by converting the energy in energy rich bonds of organic molecules to high energy phosphate bonds in adensosine triphosphate (ATP), which acts as an immediate energy source for biochemical reactions. In *oxidative phosphorylation*, carbohydrate is completely oxidised to carbon dioxide and water, yielding the maximum number of moles of ATP synthesised by the mitochondrial electron transport system. In *substrate level phosphorylation*, ATP is synthesised directly from the metabolism of the substrate and independently of the electron transport system. Substrate level phosphorylation is much less efficient than oxidative phosphorylation, yielding

Topic 12 Redox potential

only 2 moles of ATP per mole of glucose compared to 38 moles of ATP for oxidative phosphorylation.

- *Strict (obligate) aerobes* require oxygen for respiration (oxygen is used as the final acceptor in the electron transport system)
- *Facultative anaerobes* respire aerobically in the presence of oxygen but in the absence of oxygen are able to respire anaerobically
- *Strict (obligate) anaerobes* can only respire anaerobically in the absence of oxygen (oxygen is toxic).

Onset Of Reduction

Reduction is caused by a decrease in oxygen concentration in an environmental system; for example, by waterlogging in soils and sediments, or by poor aeration of water. The rate at which oxygen diffuses through water is a factor of 10^4 slower than through air. If oxygen is not replenished quickly enough, what is there will be rapidly consumed by micro-organisms and oxygen concentrations will fall. This triggers a change in the microbial population from aerobic organisms, first to facultative anaerobes, and then to obligate anaerobes. As the microbial population changes, their use of alternative electron sources to oxygen causes a series of important redox changes in environmental systems (Table 12.1).

As each of these reduction processes occurs the E_h is buffered, or *poised*, at a particular value (see Table 12.2). So, for example, while there is nitrate present the system will be poised at around +0.220 V; once the nitrate has been used up reduction of Mn(IV) will poise the system at around +0.200 V, and so on. E_h in oxidised systems is difficult to measure because there is no specific reaction controlling it. Generally, when oxygen is present the E_h is greater than +0.300 V.

The phase changes that occur are important: the gaseous products can be lost from the soil, sediment or water to the atmosphere; Mn and Fe become soluble and can be moved within or out of the system.

Table 12.1 Sequential reduction reactions with onset of anaerobic conditions.

Sequential reduction reactions		
$2NO_3^- + 12H^+ + 10e^-$ solution	$\longleftrightarrow$	$N_2 + 6H_2O$ gas
$MnO_2 + 4H^+ + 2e^-$ solid	$\longleftrightarrow$	$Mn^{2+} + 2H_2O$ solution
$Fe(OH)_3 + 3H^+ + e^-$ solid	$\longleftrightarrow$	$Fe^{2+} + 3H_2O$ solution
$SO_4^{2-} + 10H^+ + 8e^-$ solution	$\longleftrightarrow$	$H_2S + 4H_2O$ gas
$CO_2 + 8H^+ + 8e^-$ gas	$\longleftrightarrow$	$CH_4 + 2H_2O$ gas

Table 12.2 Approximate E_h values for redox reactions.

Approximate E_h values (V)	
Aerated system	> +0.300
Nitrate reduction	+0.220
Mn(IV) reduction	+0.200
Fe(III) reduction	+0.120
Sulfate reduction	−0.100
Methane formation	−0.200

Anaerobic Respiration

Anaerobic organisms respire using either substrate level phosphorylation (fermentation) which, although inefficient, does not require oxygen or oxidative phosorylation using alternative acceptors in place of oxygen (NO_3^-, MnO_2, $Fe(OH)_3$, SO_4^{2-} and CO_2) for the final step in the electron transport system. Nitrate, manganese oxides and iron oxides are used by facultative anaerobes, and sulfate and carbon dioxide by obligate anaerobes.

Aside from the inability to use the mitochondrial electron transport system to generate ATP, part of the reason for the energy inefficiency of fermentation is that the products contain much of the original bond energy. Fermentation is utilised by both facultative and obligate anaerobes producing, for example, ethyl alcohol (yeast is a facultative anaerobe) or a variety of organic acids such as formic, acetic, propionic, butyric and lactic (*Lactobacillus* is an obligate anaerobe). In addition to the products of carbohydrate metabolism, anaerobic metabolism of sulfur and nitrogen containing molecules, such as amino acids, produces a variety of partially decomposed products including thiols, such as methane thiol (CH_3HS) and dimethyl sulfide [$(CH_3)_2S$], and amines, putrescine [$NH_2-(CH_2)_4-NH_2$] and cadaverine [$NH_2-(CH_2)_5-N$]. These include some very unpleasant smelling compounds which are responsible for the 'rotten eggs' and 'bad drains' odours associated with anaerobic conditions.

Significance of Reduction in Environmental Systems

Reduction of NO_3^-, Mn(IV) and Fe(III) can occur under moderately reduced conditions, such as found in intermittently waterlogged soil or sediment.

- Reduction of nitrate causes the loss of a major nutrient. Often the reduction does not proceed to the formation of N_2 gas, and nitrous oxide (N_2O) is formed, which is a potent greenhouse gas and can cause ozone depletion in the atmosphere (see Topics 49 and 51).
- Redox changes of Mn and Fe bring about gleying in soil (see Topic 15). In sediments, the redox front can often be indicated by a sharp change in Mn and Fe concentrations at a specific depth (see Topic 18). Re-oxidation to Mn(IV) and Fe(III) results in precipitation of highly reactive oxides, which can sorb significant amounts of anions, such as phosphate and heavy metals (see Topic 8).

Sulfide and methane production occur only under highly reduced conditions, such as in permanently waterlogged sediments and peat soils:

- Sulfide may be lost as hydrogen sulfide gas, but often metal sulfides (e.g. iron sulfide, FeS) are precipitated
- Methane is a potent greenhouse gas (see Topic 49).

SECTION B

Soil

13. Soil Development

Soil is formed by the interaction of five soil forming factors:

(1) parent material
(2) climate
(3) topography
(4) time
(5) biotic factors – may be subdivided into a sixth factor, human activity.

Parent Material

Parent material is the material acted upon by physical, chemical and biological processes, ultimately to produce soil. It is usually inorganic (mineral) and is the main source of the mineral component of a soil. Parent material can be consolidated rock or unconsolidated superficial deposits:

- *Consolidated rock* (see Topic 1 for a description of the main types of rocks)
- *Unconsolidated superficial deposits* (see Topic 2 for a description of these deposits).

Parent material strongly affects two important properties of soil:

- *Soil texture* (see Topic 16) – large grain rocks (e.g. granite, sandstone) produce coarse, sandy soils; small grain rocks (e.g. basalt, shale) produce fine, clay soils
- *Inherent soil nutrient status* – rocks containing relatively high amounts of Si, K, Al and Na in minerals such as quartz and feldspar (e.g. granite and sandstone) produce nutrient-poor soil; rocks containing relatively high amounts of Ca, Mg, Fe and trace elements in ferromagnesian minerals such as olivine and pyroxene (e.g. basalt and shale) produce nutrient-rich soil.

While the nutrient status of a soil can be altered by use of fertilisers (see Topic 38), it is almost impossible to change the texture of a soil in any significant way.

Climate

The main climatic factors affecting soil formation are:

- *precipitation (rainfall and snow)*, which determines the flow of water through a soil
- *temperature*, which affects the rates of chemical reactions, water loss and biological processes in a soil.

Water movement through soil can be downward, lateral or upward. The rate at which water moves down through the soil profile depends on:

- the amount of precipitation
- the permeability of the soil
- upward movement resulting from capillarity, evaporation or transpiration.

In arid regions (low rainfall, high temperature) there are small amounts of precipitation and high rates of evaporation, so little water moves down through soil, resulting in the accumulation of weathering products and formation of saline or alkaline soils. In tropical regions (high rainfall, high temperature) there is a high amount of precipitation, resulting in intense leaching and advanced weathering, causing loss of the more readily weathered minerals and accumulation of resistant products such as hydrous oxides and kaolinite, and formation of variably charged soils such as oxisols. In temperate regions (moderate rainfall and temperature) the type of soil formed depends on the interaction of climatic factors with topography and parent material. Moderate leaching of more basic parent materials (e.g. basalt, shale) results in the formation of brown earth soils, while more intense leaching of coarse, acidic parent material (e.g. granite, sandstone) results in the formation of podzols, both of which are freely draining. If downward movement of water is impeded, e.g. by high clay texture or by fragipan, a surfacewater gley is formed; if the groundwater table moves up into the soil profile, a groundwater gley is formed. Waterlogging leading to the build up of more than 40 cm of organic matter results in formation of a peat (see Topic 15).

Topography

Topography, or the shape of the landscape, interacts with climatic factors, to determine in particular the way in which water passes through the soil profile. Rain falling on to flat ground can all move downward through the soil. On sloping ground, water can move laterally, either on the surface or within the soil profile, resulting in wetter soils at the base of slopes (Figure 13.1).

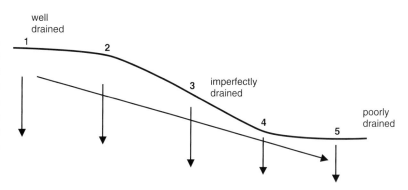

Figure 13.1   Effect of topography on water movement in soil.

Time

Soils are dynamic systems which are constantly being acted upon by formation processes. Time can affect this process in two ways. One of the other soil forming factors may change with time: a good example is climate change, which may mean soils become wetter or dryer, or are subjected to different temperatures, over time. The extent of soil formation also depends on time. In many temperate regions the parent materials from which soils are developing were exposed after the retreat of the last Pleistocene ice sheets, approximately 10 000–20 000 years ago. Thus the properties of these soils are determined by the extent to which the minerals in the parent material have weathered. Many tropical and subtropical regions were not glaciated at that time and have been subjected to much longer periods of weathering ($>100 000$ years).

Biotic Factors

This effect is brought about by living organisms in soil, which can be divided into three broad groups.

Soil Microrganisms – Viruses, Bacteria, Actinomycetes, Blue-Green Algae, Fungi, Algae (See Topic 26)

Heterotrophic organisms utilise carbon from plant and animal remains, and so are important factors in the breakdown and turnover of organic matter in soil.

Autotrophic organisms utilise specific substrates and bring about chemical transformations, e.g. nitrification (oxidation of NH_4^+ to NO_2^- and NO_3^-), oxidation and reduction of iron and sulfur species.

Exudates, mainly organic acids and chelates, released by the micro-organisms can accelerate the weathering of soil minerals (see Topic 10).

Soil Animals – Protozoa, Nematodes, Arthropods, Earthworms, Gastropods, Mammals

Soil animals are involved mainly with decomposition processes and mixing of soil. Insects and gastropods, for example, are often the first to break down plant debris, while earthworms and burrowing mammals are effective at incorporating the organic and mineral components and mixing soil vertically within the profile.

Vegetation

Plants can affect soil in a number of ways. They are the main sources of organic matter in a soil (see Topic 5), and the type of vegetation strongly influences the properties of the organic material (e.g. grass/trees, etc.). The soil immediately around the plant roots (the *rhizosphere*) is affected not only by the roots themselves, but also by the micro-organisms that proliferate there, feeding off dead plant cells and plant exudates. Root and microbial respiration has the effect of acidifying the soil in the rhizosphere. Plants can also have a physical effect on soil by promoting and stabilising cracks and channels as a result of root growth, and by aggregating the soil (see Topic 17). In addition, a plant cover helps to protect the soil surface, especially from the erosive effects of wind and water.

Human Activity

Humans have had a considerable influence on soils, especially as a result of agriculture, industry and building. Cultivation of soil for agriculture means that the natural vegetation has been removed: for example, in the UK most of the woodland cover, which would be the natural climax vegetation, has disappeared and been replaced by arable cropping or grazing. Agricultural practice means that many soils are maintained at high nutrient status, and often at a higher pH than would naturally exist, and that a whole range of artificial agrochemicals are added to soils. All of this affects the physical, chemical and biological properties of soil. Since the start of the industrial revolution, many soils have received inputs of pollutants by aerial deposition or from the deliberate addition of liquid or solid wastes, the main effect of which is on the biological activity of a soil. Buildings, roads and other developments have resulted in the covering over of many soils.

14. Soil Horizons and Soil Profiles

Soils are formed as exposed parent material weathers from the surface, with the weathering front getting deeper with time and, especially once vegetation has established, incorporation of organic material at the surface. As weathering proceeds and organic matter is added, different layers or *horizons* can be identified in soil due to such things as changes in colour, texture or water content. The *soil profile*, the vertical face from the surface to the depth where unaltered parent material is encountered, is made up of a number of soil horizons (Figure 14.1). The soil type depends on the processes that occur and the horizons formed (see Topic 15).

Horizon Nomenclature

Horizons can be either organic or mineral, and are described using a specific nomenclature system. This system varies slightly in different countries; the one described here is used in the UK.

Litter Layers and Organic Horizons

(>30% organic matter with >50% clay in mineral fraction; >20% organic matter with 0% clay in mineral fraction; for intermediate clay contents the organic matter content is varied proportionately)

- **L** Layer of surface litter – loose and relatively undecomposed, e.g. fresh leaves
- **F** Fermentation layer – partially decomposed litter with recognisable plant remains
- **H** Humus layer – well decomposed litter with no recognisable remains.

F and H horizons contain organic material in the first stages of degradation and decomposition, which accumulates under dry conditions as a result of a decline in the rate of biological activity caused by acidity or low temperature.

- **O** Organic horizon developed under wet conditions due to a decline in the rate of biological activity

Mineral Horizons

(**Humose mineral horizon:** 12–30% organic matter with >50% clay; 8–20% organic matter with 0% clay; and proportionate organic matter contents for 0–50% clay. **Mineral horizon:** <8–12% organic matter depending on clay content).

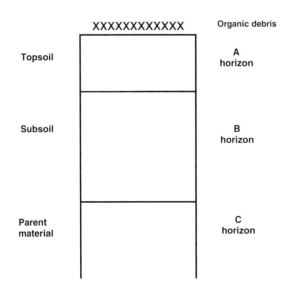

Figure 14.1 Soil profile.

Topic 14 Soil horizons and soil profiles

- **A** Mineral horizon at or near the surface, characterised by incorporation of humified organic matter and/or cultivation. Organic matter is intimately mixed with mineral material and cannot easily be separated
  - **Ap** ploughed layer – mixing due to cultivation, often with a well defined lower boundary
  - **Ah** uncultivated layer with at least 1% organic matter, well decomposed with no recognisable remains
- **E** Subsurface mineral horizon that is light in colour due to loss of organic matter and/or hydrous oxides and/or clay by eluviation into underlying horizons. It is differentiated from an O or A horizon above by its lighter colour and lower organic matter content, and from a B horizon (see below) by colour or lower clay content
  - **Ea** coatings of organic matter and iron oxide have been lost from the coarse particles, so the dominant colour is determined by the main minerals in the sand and silt fractions; these are usually quartz or feldspars, so the colour is grey-white – found above a Bh or Bs horizon and is diagnostic for a *podzol*
  - **Eb** has sufficient content of iron oxide to give a predominantly brown colour – found above a Bt horizon and is diagnostic of an *argillic soil*
- **B** Subsurface mineral horizon characterised by one or more of the following: (i) illuvial deposition of organic matter and/or hydrous oxides and/or clay; (ii) pronounced weathering and alteration of the parent material; or (iii) marked development of soil structure
  - **Bw** weathering of parent material is the dominant process, usually with obvious structural development – found in *brown earth* soils
  - **Bs** an orange-red horizon due to accumulation of deposited hydrous oxides (especially iron oxides) – found in *podzols*
  - **Bh** a dark brown-black horizon due to accumulation of deposited organic matter – found in *podzols*
  - **Bg** a mottled horizon with orange and grey colours due to redox changes of iron – found in *gley soils*
  - **Bt** contains accumulated clay that has been transported down the profile – found below an Eb or Ap horizon, and is diagnostic of an *argillic soil*
  - **Bf** a *thin iron pan* $\leq$ 1 cm thick: a red-brown to black, brittle or cemented B horizon enriched in iron, carbon and often aluminium – found between an overlying gleyed horizon (Ahg, Bhg or Eag) and underlying ungleyed B horizon
  - **Bx** a *fragipan*: a dense, uncemented horizon – usually found in soils developing on Pleistocene glacial deposits and thought to have been formed by compaction due to ice; often causes waterlogging in the horizons above
- **C** Unconsolidated or weakly consolidated mineral horizon that still shows the structure of the parent material rock, and which lacks the diagnostic criteria for A, E or B horizons described above
- **G** Grey-green horizon found at the base of a profile: the colour is due to reduction of iron as a result of the groundwater table being sufficiently high to be within the soil profile – found only in *groundwater gleys*.

Note that the suffixes g and x have been used above to describe B horizons, but can also be found in other types of horizon, e.g. Eag, Cx. These descriptors can be combined if more than one process is thought to occur in a horizon. For example, a horizon with accumulated clay, which is also compacted and gleyed would be a Btgx.

If a horizon has different properties at different depths, but these differences do not merit classification as a different type of horizon, numerical suffixes are used. For example, the Bs of a podzol may have slightly different colour lower in the horizon and so could be split into Bs1 and Bs2.

When classifying a soil, the first step is to identify the component horizons, and then to deduce the soil forming processes that are operating. The horizons make up the soil profile (see Topic 15).

15. Common Soil Types

The soils described here are the major soil types found in the UK, and are common throughout the temperate regions of the world. They are named according to the UK Soil Survey systems, with the equivalent US terminology in brackets. The soils are classified according to the dominant soil forming process (*in situ* weathering, podzolisation, gleying or peat formation) and represent the model types in each case. In the field, it is common to find soils in which it is not possible to decide what is the dominant soil forming process. Two or more processes may be occurring simultaneously, so the soil may be an intergrade between two model types. Other factors may also have an effect; for example, a strongly coloured parent material may mask the characteristic colours of diagnostic horizons.

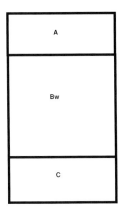

Figure 15.1 Brown earth profile.

Brown Earths (Alfisols)

The main soil forming processes are:

(1) *in situ* weathering of parent material
(2) formation of a *mull humus*.

Brown earths (Figure 15.1) are characterised by:

(1) Thin L horizon of fresh litter
(2) Well developed A horizon, containing humified organic matter intimately mixed with the mineral material
(3) Bw horizon of maximum *in situ* weathering.

Different types of brown earths may not be obvious in the field by examination of the profile morphology, but can be distinguished by chemical analysis of the soil:

- Brown earth of high base status – neutral or slightly acid pH; high exchangeable Ca, Mg, K, Na; high % base saturation;
- Brown earth of low base status – acid to slightly acid pH; low exchangeable Ca, Mg, K, Na; low % base saturation.

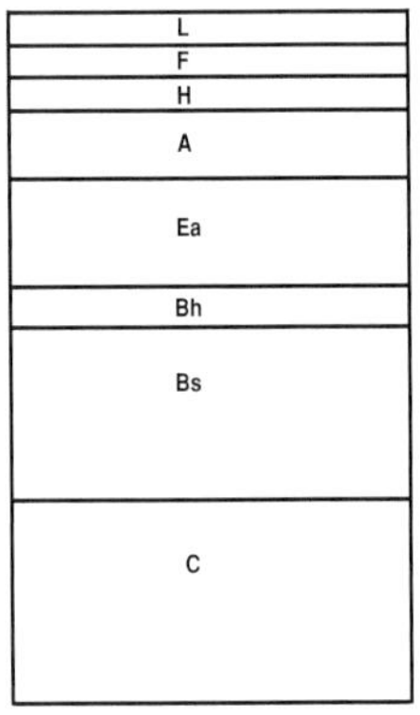

Podzols (Spodosols)

The main soil forming processes are:

(1) mobilisation leaching and redeposition of organic matter, iron oxides and aluminium oxides
(2) accumulation of *mor humus* under acid conditions.

Podzols (Figure 15.2) are characterised by:

(1) Ea horizon, which is grey–white in colour due to eluviation of organic matter and/or hydrous oxides (iron oxide)
(2) Bs horizon, which is orange-red in colour due to the illuvial deposition of hydrous oxides (iron oxide)
(3) In an iron-humus podzol, a Bh horizon in which organic matter has been redeposited
(4) Mor humus accumulating at the surface comprising L, F and H horizons
(5) The organic matter in the A horizon has been washed down from the mor humus and is present as coatings on the mineral grains.

Figure 15.2 Podzol profile.

Depending on the extent to which the various soil forming processes have occurred, a number of different types of podzols may be recognised:

- *Iron-humus podzol* – movement of both hydrous oxides and organic matter (both Bs and Bh horizons in profile)

Topic 15 Common soil types

- *Iron podzol* – no Bh horizon in the profile
- *Iron pan podzol* – formation of a thin, hard iron pan (Bf) due to changes in the redox potential at the Bh/Bs interface, impedes downward movement of water
- *Peaty podzol* – the wetness which develops above the iron pan eventually results in an accumulation of organic matter forming a peaty surface O horizon.

Gleys

The main soil forming process in gley soils are:

(1) Reduction and reoxidation of iron and manganese oxides due to changes in aeration resulting from temporary or permanent waterlogging
(2) There may also be organic matter accumulation at the surface depending on the degree of wetness.

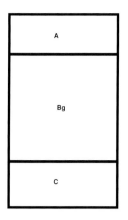

Figure 15.3 Surface water gley profile.

Two types of gleys may be recognised:

- *Surface water gleys (Figure 15.3)*
  - Reduction results from impeded drainage of water through the soil due to factors such as a high clay content or the presence of a pan
  - The characteristic horizon is a Bg, in which grey mottles due to reduced iron compounds are found on ped faces and along pores; during extensive dry periods reoxidation may occur resulting in orange mottles
- *Ground water gleys (Figure 15.4)*
  - Waterlogging is due to the presence of a groundwater table within the profile
  - The diagnostic horizon is a G, which is permanently waterlogged and has a grey-green colour
  - Above this there is a Bg in which oxidised, orange mottles are associated with the ped faces and the pores.

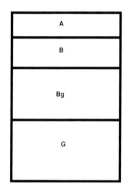

Figure 15.4 Groundwater gley profile.

If an organic O horizon of 7.5–40 cm develops on the surface due to wetness, the soil is called *peaty gley*.

Peats (Histosols)

The main soil forming process is the accumulation of organic matter at the surface under cold, wet conditions.

To be classified as a peat, the O horizon must be greater than 40 cm deep (Figure 15.5). It may be made up of various types of O horizon, distinguished by the degree of humification of the organic matter. Any processes in the underlying soil are not considered to be important. If the O horizon is between 7.5 and 40 cm deep, the term *peaty* is used to qualify the other dominant process occurring below the O horizon, e.g. *peaty gley*, *peaty podzol*, etc.

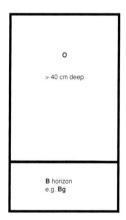

Figure 15.5 Peat profile.

16. Soil Texture

Soil texture is the size distribution of the mineral particles in a soil. There is a conventional arbitrary division at 2 mm particle diameter, with the larger material, stones and gravel, usually discarded. This fraction is not normally used in soil analysis, but it can give some information about the soil, such as aeration status and the origin of the parent material. The smaller than 2 mm, or *fine earth*, fraction is subdivided into three broad groups based on particle diameter: sand sized particles, silt sized particles and clay sized particles. (Note that the term 'clay' is used in different contexts. In Topic 3 the clay minerals were discussed, but here we are dealing with a fraction of the soil based on particle size, which contains not only the clay minerals but other minerals such as oxides and carbonates.)

Textural Size Fractions

Slightly different classification schemes for defining groups of different particle size are used across the world. The one given here is based on the system used by the Soil Survey of England and Wales (Table 16.1).

Table 16.1 Classification of textural size fractions.

Soil fraction	Equivalent settling diameter (mm)
Sand	
Coarse	2–0.2
Fine	0.2–0.06
Silt	0.06–0.002
Clay	<0.002

Particle Size Analysis

To measure the distribution of particle size for textural analysis, it is necessary first to disperse the soil particles by destroying any aggregating agents binding them together, and then to separate the different size fractions.

Dispersion – Breaking Down Soil to Its Primary Particles

Various methods are used, often in combination, to destroy aggregating agents and break any bonds between mineral particles:

- *Acid* – dissolves carbonate and hydrous oxides
- *Hydrogen peroxide (H_2O_2)* – oxidises organic matter
- *Alkali (sodium hexametaphosphate)* – increases negative charge on particles causing repulsion and replaces multivalent cations such as Ca^{2+} with monovalent Na^+
- *Ultrasound* – physical dispersion of particles.

Fractionation by Particle Size

The sand fractions are separated by sieving. Coarse sand can be trapped in a 100 mesh (150 μm) sieve and fine sand in a 300 mesh (53 μm) sieve. The sieves are placed one on top of the other and the dispersed soil suspension washed through. The two sand fractions can then be dried and weighed.

Particles smaller than 0.06 mm (60 μm) are washed through sieves into a 1 litre cylinder and fractionated by sedimentation using Stokes' law:

$$v = \frac{g(\rho_p - \rho_w)d^2}{18\eta}$$

where v is the settling velocity (cm/s), g is the acceleration due to gravity (981 cm/s^2), ρ_p is the density of the particle (2.6 g/cm^3), ρ_w is the density of water (0.998 g/cm^3 at 20°C), η is the viscosity of water (0.010 g/s/cm at 20°C) and d is the equivalent diameter of spherical particle (cm).

The distance through which particles of a known size will fall in a given time can be calculated. The suspension is shaken and allowed to settle for that time; then a known volume taken out by pipette from the required depth is dried and weighed. Two major assumptions in this method are that all particles have a density of 2.6 g/cm^3 and that the soil particles are spherical. Neither of these is true, and so some error is introduced into the measurement.

Once the samples have been dried the different size fractions can be weighed, giving values for sand (coarse + fine), silt and clay. These can then be expressed as a percentage of the total sample and plotted on a triangular textural diagram (Figure 16.1) to give the textural class of the soil. The texture affects the feel of a soil: sandy soils are gritty, silty soils are silky and clay soils are sticky, but organic matter will modify these descriptors.

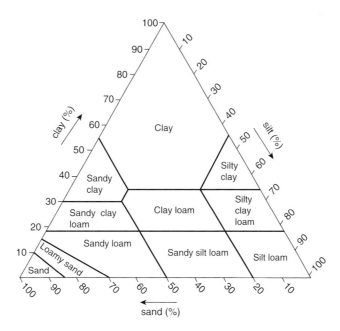

Figure 16.1 Soil texture classification.

Soil Properties Affected by Texture

The texture particularly influences the pore size distribution in a soil. Larger particles cannot pack together as closely as fine particles, so sandy soils have a greater number of large pores compared to clay soils, which tend to have small pores. Thus the soil properties most affected by texture are those related to air and water movement: permeability, water retention, leaching and aeration. Also affected are root penetration, as roots tend to grow into soil pores, and soil temperature, as air heats up at a faster rate than water.

Sandy soils have good drainage, are easy to cultivate, warm up quickly and allow good root penetration. On the negative side, they have poor water holding ability, poor nutrient supply and readily lose nutrients by leaching.

Clay soils, on the other hand, have good water holding ability, good nutrient supply and can retain nutrients. Conversely, they are poorly drained and easily become waterlogged, are easily damaged by animals or machinery, warm up slowly, become hard when dry and have limitations to cultivation.

For agricultural purposes, soils with a loamy texture, lying towards the middle of the triangular diagram, are best. They have a mixture of particle sizes and retain the good points but limit bad points of very sandy or clay soils.

17. Soil Structure and Aggregation

Soil structure describes the aggregation of sand, silt, clay and organic matter into units called *peds*. In particular, it describes the shape and size of the peds. The aggregating bonds between different soil particles are essentially due to the processes described in Section A: flocculation (Topic 6), ion exchange (Topic 7) and adsorption (Topic 8). In addition, some soil components, such as carbonate, oxides and carbohydrate, may act as cements physically holding soil particles together, and plant roots and fungi can physically aggregate soil particles.

Peds are separated from each other by surfaces of weakness, so structure also defines the pore system in a soil. Therefore soil structure affects similar properties to soil texture: movement of water, air, roots and micro organisms. Whereas soil texture is an inherent property of a soil, derived from the parent material, which cannot be substantially affected by soil management, soil structure can be affected by soil management. Factors such as additions of organic matter, additions of lime, cultivation of soil and the choice of crop can all affect soil structure.

Classification of Soil Structure

Two broad classes can be identified.

(1) *Structureless soils* (apedal soils) have no obvious aggregation.
   - *Single grained* – no aggregation of primary particles, mainly individual sand sized particles. Typical of very sandy soils, such as desert sands and sand dunes
   - *Massive* – uniformly cohesive soil with no natural planes of weakness. Typical of some very clay subsoils.
(2) *Structured soils* (pedal soils) show aggregation into peds of different shapes (see Figure 17.1).
   - *Spheroidal*. Irregular peds with all three axes equally developed and adjacent faces that do not fit together. Two types are recognised: *crumbs*, which are porous, and *granular*, which are denser and less porous. They are typically found in well structured A horizons under grass and forests. The larger inter-ped pores allow free drainage of water, while the smaller intra-ped pores allow retention of plant available water. They provide a good, well-aerated environment for roots and micro-organisms
   - *Platey*. Plate-like peds, with the vertical axis much smaller than the horizontal axis. They are often found at the base of an A horizon above an impermeable B horizon or in surface due to compaction. They impede downward movement of air and water, and can lead to anaerobic conditions in the surface layers
   - *Prismatic*. Prism-like peds, with the vertical axis much larger than the horizontal axis. They are typically found in Bw and Bg horizons. They allow movement of air and water down profile, although the peds may fit tightly together when soils are wet, but shrink with opening of pores when soil dries. *Columnar* peds are similar to prismatic peds, but their tops are rounded due to breakdown of structure and washing off of soil particles into pores. This can impede drainage of water
   - *Blocky*. Block-like peds, with all three axes equally developed as in spheroidal peds, but in this case the adjacent faces fit well together. They allow good movement of air and water, and good root penetration. Typically found in lower A and upper B horizons.

Aggregation of Soil

A number of factors act to promote the aggregation of soil particles:

(1) Wetting–drying cycles can create pressure on soil particles due to swelling and shrinking which help to break down large aggregates. The presence of dissolved salts can add to this effect by creating an osmotic pressure. Drying helps to stabilise aggregates
(2) Freeze–thaw cycles too can create pressure as ice expands when formed, again helping to break down larger aggregates
(3) Plant roots aid aggregation in a number of ways. They tend to grow into existing pores and the pressure exerted by the root widens the pore and root exudates and dead cells act to stabilise the pore wall. The exudates also act as cements, holding soil particles together, at least for short periods of time. Roots systems can physically bind aggregates together
(4) Microbial and faunal activity also aid aggregation. Micro-organisms too produce exudates that act as cements. They also breakdown plant residues in the humification process, and the humic material produced provides long term stability to aggregates. Fungal hyphae act in the same way as roots to physically bind aggregates. Soil fauna, especially invertebrates such as earthworms and mites, ingest large amounts of soil and their faeces form very stable microaggregates.

Topic 17 Soil structure and aggregation

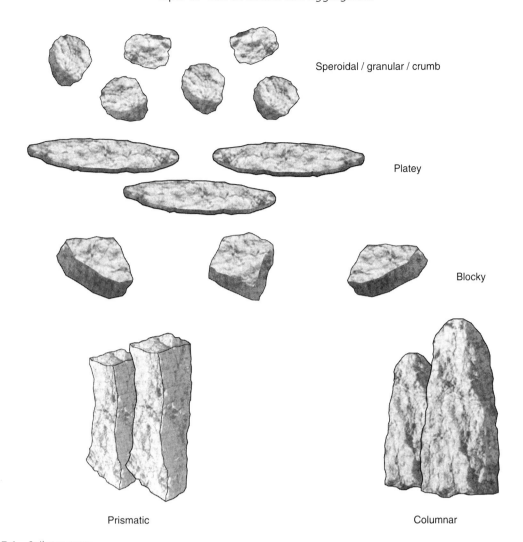

Figure 17.1 Soil structures.

Soil aggregation is usually considered on a number of levels. The smallest structural units are *clay domains* (see Topic 6), which are flocculated clay unit layers that act as an entity and have stability when wet. They may be a few nanometers thick and up to about 5 μm in their other axes.

Microaggregates are units less than 250 μm in diameter made up of clay domains, sand and silt particles held together by humified organic matter and carbohydrate (Figure 17.2). In some soils, edge-to-face flocculation of kaolinite and bridging by hydrous oxides may also be important (see Topic 6).

Macroaggregates are units greater than 250 μm in diameter, made up of microaggregates held together by plant roots, fungal hyphae and other cementing agents such as hydrous oxides.

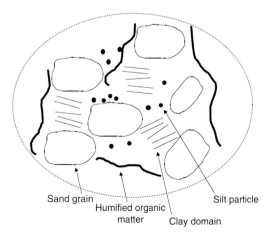

Figure 17.2 Microaggregate structure.

Sediments

18. Sediments – Processes

Sediments are formed from inorganic, or sometimes organic, material that has been suspended in, transported by and deposited from water. They consist of the solid material and the *porewater* (or *interstitial water*), and many are overlain by water. Some sediments are transported by wind (aeolian sediments). Sediments are important components of the natural geochemical cycle, i.e. weathered material that is moved around various parts of the environment, but are especially important in the hydrological cycle. They are a major reservoir for contaminants, and because of the normally long residence times are effectively permanent sinks for some. Because of the way in which they are deposited in successive layers, sediments can provide a historical record of environmental change (e.g. climate change, different vegetation types and contaminant inputs).

Three broad categories of sediment can be identified, depending on the origin of the solid component:

- *Detrital* (sometimes called clastic or terrigenous) – derived from the weathering and transport of rock particles or soil components. They comprise minerals such as quartz and feldspars, which persist as primary minerals in soils, and the clay minerals, formed as secondary minerals in soils. These are eroded from soils and released by direct weathering of rocks (see Topics 1, 2 and 3)
- *Authigenic* (chemical) – derived from the products of *in situ* chemical reactions within a sediment or the associated water. Examples of authigenic minerals are hydrous oxides, especially of Fe and Mn which both readily undergo redox changes in the sediment environment, and pyrite (FeS_2), which contains reduced forms of Fe and S and is commonly found in highly reduced sediments.
- *Biogenic* – derived from the waste products and remains of living organisms. This may be the organic remains of organisms that persist under anaerobic conditions, or materials such as calcite ($CaCO_3$), apatite ($Ca_5(PO_4)_3(OH,F)$) and silica ($SiO_2.nH_2O$), which are used by some organisms as structural materials such as shells and skeletons.

As can be seen from the above, sediments are chemically similar to soils because they are formed mainly from the weathering products of rocks with an input of secondary minerals. They too consist of a solid phase in contact with water, although in the case of many sediments, water usually makes up a greater fraction of the total weight compared to soil. The porewater can constitute more than 90% by weight in near surface sediments; at greater depth, the weight of sediment exerts increasing pressure, leading to loss of porewater and compaction of the sediment. Over geological timescales this results in the formation of the sedimentary rocks.

Diagenetic Processes in Sediments

The term *diagenesis* refers to the various interacting processes that occur in a sediment after deposition. As in soils, a mix of physical, chemical and biological processes affect the properties of the sediment.

Physical Processes

The main effect is that of compaction, resulting in a greater density and lower water content at depth. In addition, there can be some movement and sorting of surface sediment due to resuspension by the action of water currents.

Chemical Processes

Sediments tend to be reduced environments due to the inputs of organic matter and the high water (and hence low air) content. The same redox reactions described for soils in Topic 12 also occur in sediments. Nitrate reduction is less important than in soil, but redox changes in Fe and Mn are major processes, especially where the sediment undergoes cycles of oxidation and reduction. In highly reduced sediments sulfide and methane production are important.

In the oxidised layer, the depth of which will vary according to the position of the saturated zone within the sediment, iron exists as Fe(III) and manganese as Mn(IV), both of which are insoluble. Below this layer the redox potential falls with depth, and Mn(IV) is reduced to the soluble Mn(II) and, further down at lower Eh, Fe(III) to Fe(II). Once these ions have been released into solution there is a tendency for them to diffuse upward into the porewater in the oxidised layer, where

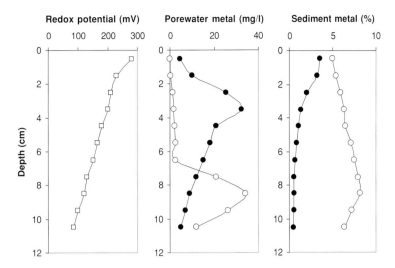

Figure 18.1 Fe and Mn distribution in a sediment profile (-○- Fe, -●- Mn).

concentrations are much lower, where they are reoxidised. This generates a sediment profile distribution of a near surface maximum concentration of Mn in the solid phase with a porewater maximum slightly below. The same pattern occurs for iron, but at a lower depth because the redox change for iron occurs at a lower Eh value. This pattern is shown in Figure 18.1. Although only a minor component of most sediments on a weight basis, these oxides are extremely important as they provide highly active surfaces on to which other ions can be adsorbed.

Sulfide formation occurs in highly reduced sediments, iron sulfides being especially important. Initially, FeS is formed, which is a black sediment often found, for example, immediately below the orange, oxidised surface layer of recently exposed lake sediments, or at depth in beach sand. The long term stable form is pyrite (FeS_2), which is commonly found in coal and sedimentary rock such as shale which has a high organic content. If such reduced sediment is exposed to aerobic conditions, the oxidation of the Fe and S results in acidification – found, for example, in some reclaimed river deltas and in polders reclaimed from the sea. The same process occurs in some coal mine wastes (see Topic 46).

Biological Processes

Living organisms can cause considerable changes by mixing sediment particles. Benthic (or bottom dwelling) organisms bring about mixing of surface sediment by their activities, and in some cases its resuspension into the water column. Further mixing is caused by other species feeding on the benthic organisms. The extent of this effect is dependent on the density of the species populations. Where this is high, the surface sediment may be completely mixed. Organisms that live within the sediment can cause mixing of surface material to depth. Such processes must be taken into consideration when sediments are used as indicators of historical environmental change.

19. Sedimentary Environments

Sediment Transport

Sediment particles that are transported by water can be separated into different size fractions depending on the size of the particle, its density and the velocity of the water flow (current energy). Sedimentation is determined by Stokes' law (see Topic 16), which relates the settling velocity of a particle to its diameter and density. Thus large, heavy particles will sediment out of moving water more readily than small, light particles. The flow of water is important: sediment is retained in suspension for longer in fast moving water.

The flow of water over a sediment will bring about some resuspension of particles, with small, light particles being taken up by slower moving water (lower current energy) than large, heavy particles. This effect is complicated by the cohesiveness of the sediment. Clay and silt rich sediments are more cohesive than sandy sediments, and therefore more difficult to bring back into suspension.

Overall, the effect of these processes is to concentrate coarse, sandy sediments in fast flowing (high current energy) water, and fine, silt and clay sediments in slow flowing (low current energy) water.

When sediments are transported by wind, the smaller particles are more easily moved, and transported over greater distances.

River Sediments

The mineralogical makeup of river sediments is dependent on the geology of the catchment. Mixing occurs due to the processes outlined above, but overall the sediment tends to be coarser in the upstream (usually faster flowing) part of the river, becoming finer downstream. Heavy minerals in river sediments are used for geochemical mapping purposes, especially to identify the presence of metal ore deposits in the catchment. Measurement of metal concentrations in river sediment deposited in floodplains can provide a record of contamination in the river system.

Lake Sediments

Lakes are fed by rivers, and when the river enters a lake there is usually a decrease in water velocity. As a result, coarse sediment is deposited close to the entry point, with finer sediment being transported further into the lake. Energy derived from waves and currents within the lake results in further sorting based on particle size. The chemical nature of lake sediment is highly dependent on the catchment area. In upland areas where peat soils predominate, lake sediments tend to be highly organic, whereas their properties in areas of mineral soils depends on the mineralogical makeup of the parent materials. Material derived from outwith the lake is termed *allochthonous*, while diagenetic material formed in the lake is called *autochthonous*.

Lake sediments typically accumulate at the rate of a few millimetres per year. As lakes act as sinks for contaminant inputs to a large area (the catchment), sediment cores can provide a good historical record of such inputs. Iron and manganese diagenesis (see Topic 18) are particularly important, providing highly reactive sorptive surfaces for ions such as phosphate, arsenate and many heavy metals.

Nearshore Marine Sediments

Much of the sediment transported by rivers to the sea is rapidly deposited close to the shore for a number of reasons. As when rivers flow into lakes, there is a decrease in water velocity in the estuary causing sediment particles to settle out of the water column. The river water tends to be relatively acidic and of low ionic strength (i.e low content of dissolved ions) compared to the high ionic strength, alkaline seawater. Such conditions favour formation of authigenic iron and manganese oxides, which not only settle out of solution themselves, but also can bind with negatively charged clay minerals and humic material to form large aggregates that readily settle. The clay particles tend to be dispersed in the river water, but flocculate (see Topic 6) in the high salt concentrations in seawater, resulting in sedimentation. Rivers also transport relatively large amounts of nutrients to the sea, so estuaries have a high biological productivity. The waste and decay products of these organisms make nearshore sediments high in organic matter.

Accumulation rates in nearshore sediments range from the order of a few millimetres to centimetres in low energy environments to almost zero in high energy environments. High energy environments can bring about considerable mixing of sediments to a significant depth – up to about 1 metre.

Continental Shelf and Slope Sediments

Away from estuaries and the influence of rivers, the supply of sediment to marine systems decreases. Such areas are transition zones between the nearshore sediments and deep sea sediments and are the most dynamic of the marine systems. Processes such as mass slumping of sediment down the continental slope and water currents result in rapid, wholesale movement of sediment, so little if any accumulation occurs.

Saltmarsh Sediments

Saltmarshes are another transition zone, this time between nearshore and land based systems. They form when vegetation, especially salt resistant grasses, colonise intertidal sediments. The vegetation slows up the incoming tidal water, causing sediment deposition. Furthermore, the grasses themselves act as physical traps, resulting in a greater build up of sediment. Saltmarsh sediment tends to be fine grained, so it can hold and retain high concentrations of contaminants.

Deep Sea Sediments

Relatively little terrigenous sediment reaches the deep sea sediments, that which does being aeolian (wind blown) material. The major input is of biogenic origin, derived from plankton, which constantly settles out of the water column ('marine snow'). Two types predominate, calcareous (mainly calcite) and siliceous (silica), coming from structural components of the marine organisms. Organic matter does not persist, being broken down by microbial action in the water. Typical accumulation rates are a few centimetres per thousand years, with mixing occurring only down to about 20 cm.

SECTION D

Water

20. The Hydrological Cycle

The *hydrosphere* comprises all the water on, or close to, the surface of the Earth. The *hydrological cycle* represents the movement of water through the various reservoirs that make up the hydrosphere and its transformations between the solid, liquid and gaseous phases.

Water plays an important role in the atmospheric energy transfer system (through evaporation, transport and condensation) and liquid water is essential to chemical reactions (such as ion exchange, solubility, redox, pH, etc.) in solution and to the presence of life on Earth. The physical properties of water (Table 20.1) mean that under the normal range of environmental conditions it can exist in all three states; solid, liquid and vapour. Liquid water has a relatively high specific heat capacity and a maximum density at 4°C. It acts as a solvent for the dissolution of solids and gases. In contrast to the solubility of solids, the solubility of gases decreases as the temperature increases.

Table 20.1 Physical properties of water.

Melting point (°C)	0
Boiling point (°C)	100
Specific heat capacity (J/K/mol)	75
Latent heat of fusion (kJ/mol)	6
Latent heat of vaporisation (kJ/mol)	41
Density	1.0
At 4°C (g/cm^3)	
Ice (g/cm^3)	0.92
Solubility of O_2 in water at 1 atm	
0°C (mg/l)	14.59
20°C (mg/l)	9.07

Definitions

Precipitation	The deposition of moisture from the atmosphere as rain, hail, snow, mist and hoar frost
Evapotranspiration	The transfer of water vapour from the Earth's surface to the atmosphere by *evaporation* from liquid water surfaces and by *transpiration*, the biological process where plants take up water from the soil and lose it through their leaves
Humidity	The partial pressure of water vapour in air
Saturation vapour pressure	The partial pressure of water vapour when air is saturated at given temperature
Relative humidity	The partial pressure of water vapour in the atmosphere expressed as a percentage of the saturation vapour pressure

The Hydrological Cycle

Water vapour in the atmosphere condenses and falls as precipitation (rain, snow, etc.). Some of this water will run off the land surface to form streams, rivers and lakes, while some will percolate through soils, subsoils and porous rock strata before emerging as surface water, some of which flows to the oceans carrying dissolved and suspended material. Evaporation from water surfaces and transpiration by plants complete the cycle by adding water vapour to the atmosphere (Figure 20.1).

The total amount of water in the hydrological cycle is estimated at 1.4×10^6 teratonne (Tt). With over 97% of this water in the oceans and a further 2% in the form of snow and ice, the remaining reservoirs combined together make up just 0.6% of the Earth's water (Table 20.2).

Estimates are also available for some of the fluxes between reservoirs. Figure 20.2 shows a net flow of water from the oceans to the land of 40 teratonne per year (Tt/year) resulting from the greater rainfall over the land compared to the oceans. This water flows to the oceans as rivers containing dissolved and suspended particulate matter.

For a well mixed reservoir at steady state it is possible to calculate the mean residence time (the average time a water molecule spends in that reservoir).

$$\text{Mean residence time} = \frac{\text{Reservoir size}}{\text{Total flux in or out}}$$

Many of the reservoirs cannot be considered to be well mixed (or even at steady state), but useful comparisons can be made using this calculation:

$$\text{Residence time oceans} = \frac{1.4 \times 10^6 \text{ Tt}}{420 \text{ Tt/year}} = 3333 \text{ years}$$

By contrast the estimated mean residence times for the atmosphere and the biosphere are 10 days and 7 days, respectively.

Topic 20 The hydrological cycle

Table 20.2 The hydrosphere reservoirs.

Reservoir	Capacity (Tt)	(% of total)
Oceans	1 400 000	97.4
Snow and ice	27 000	2.0
Groundwater	8 000	0.6
Lakes and rivers	280	0.02
Soils	70	0.005
Atmosphere	14	0.001
Biosphere	1.4	0.0001

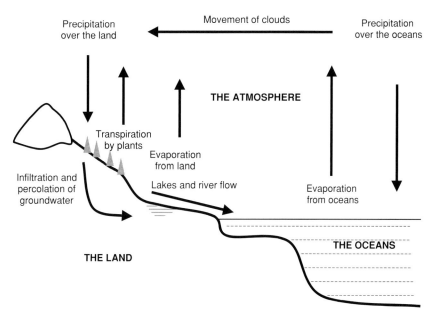

Figure 20.1 The hydrological cycle.

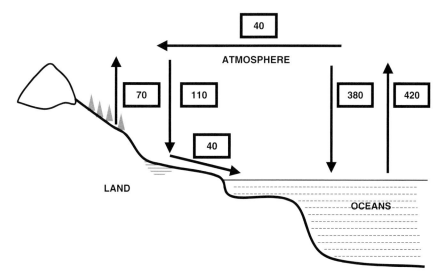

Figure 20.2 Precipitation and evaporation fluxes (in Tt/year) on land and over the oceans.

21. The Freshwater Environment

Freshwater comprises approximately 2.6% of the water in the hydrosphere and includes snow, ice, soil water, groundwater, streams, river, lakes and atmospheric water.

Precipitation

Total precipitation over the land is approximately 110 teratonne per year (Tt/year), giving an annual average rainfall of 700 mm, but with a large spatial variation. The chemical composition of rainwater varies with location due to inputs from sea spray, dust and atmospheric gases dissolving in the water droplets in clouds (Table 21.1). Near continental margins with prevailing winds from the sea, the chemical composition of the rainfall is dominated by inputs from sea spray and therefore sodium and chloride. In the continental centres, the composition is dominated by inputs from dust and thus by the soils and geology, resulting in a decreased ratio of sodium to calcium and magnesium.

The pH of unpolluted rainwater is approximately 5.7 resulting from the ionisation of dissolved carbon dioxide (see Topic 11). In industrialised areas higher concentrations of acidic gases such as SO_2, NO_2 and HCl result in lower pH values and acid rain may be of pH 4 or below (see Topic 52).

Table 21.1 Average chemical composition of rainwater.

Constituent	Concentration (mg/dm³)
Na^+	2.0
K^+	0.3
Mg^{2+}	0.3
Ca^{2+}	0.1
Cl^-	3.8
SO_4^{2-}	0.6
HCO_3^-	0.1
pH	5.7

Data taken from R.M. Garrells and F.T. Mackenzie, *Evolution of Sedimentary Rocks*, Norton, New York, 1971.

Groundwater

Groundwater is the water held in the pore system of soils and porous rock strata (Figure 21.1).

Water percolates down through the vadose zone to the water table, below which the rock pores are saturated with water. The capillary fringe immediately above the water table has a higher water content than the vadose zone due to the capillary attraction of the water for the pore system in the rock.

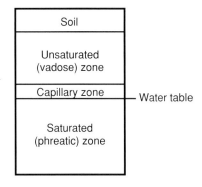

Figure 21.1 The water table and groundwater zones.

Definitions

Porosity   The fraction of the total rock that is void space providing the ability to hold water

Permeability   The ability of a rock to transmit water

Aquifer   A rock with high porosity and high permeability providing water storage and water movement. The main aquifer rocks are sandstone, limestone and chalk

Aquiclude   A rock stratum with low permeability that prevents water movement

The chemical composition of groundwater is strongly influenced by the composition of the soil and rock strata through which it has percolated and the residence time within the rock. This will influence both the overall concentration of dissolved ions and their relative abundance as shown in Table 21.2.

Table 21.2 Composition of spring water from three rock types.

	Concentration (mg/dm³)		
Ion	Chalk	Basalt	Granite
Na^+	8	6	6
K^+	<1	<1	2
Ca^{2+}	104	26	4
Mg^{2+}	1	6	3
Cl^-	15	7	9
SO_4^{2-}	12	10	6
HCO_3^-	290	99	26

Rivers and Lakes

The *river catchment* area represents the area drained by a river system, the *watershed* being the division between adjacent catchments (Figure 21.2).

Topic 21 The freshwater environment

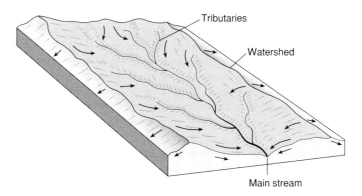

Figure 21.2 A river catchment.

Depending on the topography of the landscape, river flow (and chemical composition) will respond more rapidly or more slowly to rainfall events. This is shown by the storm hydrograph (Figure 21.3). A rapid response to rainfall is expected with steep hillsides, shallow soils, a lack of vegetation, frozen or saturated soils, all of which promote rapid run off of surface water and erosion. By contrast, shallow slopes, deep soils, good vegetation cover and dry soils all promote infiltration of water into the soil profile increasing the time interval between the rainfall and the peak river flow.

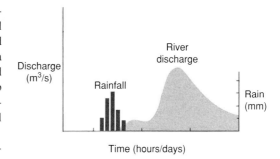

Figure 21.3 A storm hydrograph.

Because rainwater dissolves soluble weathering products as it percolates through soils and rocks, the chemical composition of rivers is strongly influenced by the local geology and generally has higher total dissolved salts and a lower Na:Ca ratio compared to rainwater. In dry areas evaporation will lead to an increase in total dissolved salts, but may also result in the precipitation of insoluble calcium salts leading to an increase in the Na:Ca ratio. In addition to dissolved material, a river also transports suspended material and a *bed load* (material rolled along the river bed) depending on the velocity of the flow.

Structure of Lakes

The water column can be divided into the *photic zone*, where carbon dioxide fixation by photosynthesis exceeds carbon dioxide production by respiration, and the *aphotic zone*, where there is overall carbon dioxide production (Figure 21.4). Light penetration may be only a few metres if the water contains much suspended material. The *benthic zone* refers to the lake bottom from the beach (*littoral zone*) to the deepest parts of the lake bed.

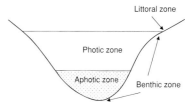

Figure 21.4 Structure of a lake.

In the summer lakes may undergo *thermal stratification* leading to the separation of warm, less dense water (the *epilimnion*) overlying colder, denser water (the *hypolimnion*), with little mixing and transfer of solutes between the layers.

22. The Marine Environment

The oceans cover approximately 70% of the Earth's surface and contain 97.6% of the water in the hydrosphere (1.4×10^6 teratonne). This environment ranges from the seashore, through the continental shelf and ocean floor to the deepest parts of the oceans, the ocean trenches, which may be over 10 km in depth (Figure 22.1). The *photic zone*, the depth to which sunlight penetrates and primary production takes places, extends to a depth of a few hundred metres depending on the clarity of the water.

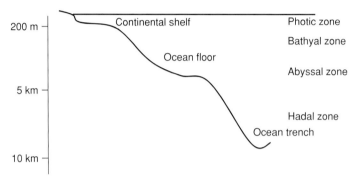

Figure 22.1 The ocean profile.

The permanent *thermocline* in the oceans results in a mixed surface layer up to 300 m deep, overlying a distinct cold layer below 1000 m.

Salinity

The average salinity of the ocean surface is 35 g/l total dissolved salts, which is usually expressed as 35 parts per thousand (35‰). Salinity varies with the ratio of inputs of fresh water and losses by evaporation, producing both global and local scale variations. Surface salinity is at a maximum in subtropical areas, corresponding to the major deserts on the land masses, where evaporation greatly exceeds rainfall, and decreases towards both the equator and the poles where rainfall exceeds evaporation. On a local scale, melting ice and large river inputs result in reduced salinity, while shallow lagoons at low latitude tend to have increased salinity due high evaporation rates. These variations are surface phenomena; below approximately 1.5 km depth the salinity is more uniformly close to 35‰.

Chemical Composition

One of the important principles of ocean chemistry is that, although the salinity of seawater varies, the ratios of the major elements remain constant, showing that it is primarily caused by evaporation and dilution; this concept is known as the *constancy of composition of seawater*.

Comparison of Tables 21.1 and 22.1 shows that seawater is not simply river water concentrated by evaporation. For example, the sodium and chloride concentrations are much too high and the calcium, bicarbonate and silicate concentrations too low, showing that chemical processes such as precipitation and adsorption are important in controlling the composition of seawater. Seawater in equilibrium with atmospheric carbon dioxide is approximately pH 8.1–8.3. The presence of dissolved salts affects the density and freezing point of water. For seawater at atmospheric pressure the maximum density (1.028 g/l) occurs at the freezing point ($-1.91°C$). The dissolved salts also affect the solubility of oxygen. For seawater at 20°C in equilibrium with air at 1 atm, the solubility of oxygen is 7.4 mg/l, compared to 9.1 mg/l for pure water.

Table 22.1 Major anions and cations in seawater.

Constituent	Concentration (mg/dm³)
Na^+	10.56
Mg^{2+}	1.27
Ca^{2+}	0.40
K^+	0.38
Sr^{2+}	0.013
Cl^-	18.98
SO_4^{2-}	2.65
HCO_3^-	0.14
Br^-	0.065
BO_3^{3-}	0.026
SiO_4^{4-}	0.002
F^-	0.001

Element Profiles

Some elements (Cl, K, Na, S) show little change in concentration with depth; these are usually present in high concentration or not involved in biological cycling and are termed *conservative elements* (Figure 22.2). Other elements show pronounced changes with depth and are termed *non-conservative elements*. Important nutrient elements N, P, Cu, Fe, Ca and Zn are depleted in the photic zone due to uptake by living organisms. Elements such as Al, Mn and Pb decline with depth due to preciptiation and adsorption reactions leading to sedimentation. Much of the adsorption occurs onto the surface of bacteria rather than inorganic particles.

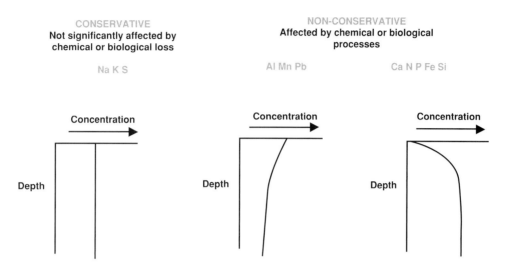

Figure 22.2 Element profiles with depth in the ocean.

The surface waters of the oceans are typically supersaturated with oxygen. This is mainly due to incorporation of air by wave action and increased dissolution at depth due to the increased pressure, but in addition within the photic zone photosynthesis contributes more oxygen than respiration consumes. Below the photic zone respiration exceeds photosynthesis; as oxygen supply from above is limited by lack of mixing and slow diffusion, oxygen concentrations fall leading to an oxygen minimum in the range 500–1000 m depth. Below this depth the oxygen concentration is often higher due to cold, oxygen rich currents flowing from the polar regions

Elemental Balances of the Oceans

The chemical composition of the oceans reflects the balance between the gains and losses of each element. The inputs from the land include rivers, glacial melt, deposition of atmospheric dust and rainfall (Table 22.2). In addition, hydrothermal activity on the ocean floor is a further source. The major input comes from the suspended load and dissolved salts in river water, amounting to 20 gigatonne per year (Gt/year). Much of this is laid down directly as sediment and it is only the elements in solution which are of interest here.

Losses include precipitation, sorption and biological uptake, coupled with sedimentation and losses to the atmosphere as aerosol droplets of sea spray, some of which is transferred to the land masses as dissolved constituents in rainfall. Comparison of the chemical composition of the lithosphere with the chemical composition of river water shows that weathering of rocks cannot account for the levels of chloride in river water. Virtually all the chloride flowing into the oceans has come from the oceans via sea spray and deposition in rainfall. This comparison also shows that the chloride in the oceans could not have originated by the accumulation of chloride weathered from rocks. The main source of the chloride present in the oceans has been hydrogen chloride gas evolved by volcanic activity over the Earth's geological history.

Table 22.2 Sources of elemental inputs to the oceans.

Rivers	90%
Suspended load	72%
In solution	18%
Glacial melt	8%
Antarctic	7%
Atmospheric dust	0.5%

Ocean Currents

One major driving force for the ocean currents is differences in density due to differences in temperature or salinity and is known as the *thermohaline circulation* (Figure 22.3). At the poles, cold, dense water sinks and spreads through the oceans beneath the permanent thermocline constrained by the land masses, eventually rising and warming to complete the circulation.

The second driving force for surface currents is the global wind pattern. Combining the effects of wind patterns and the thermohaline circulation provides a general explanation for the pattern of cold and warm surface currents (Figure 22.4).

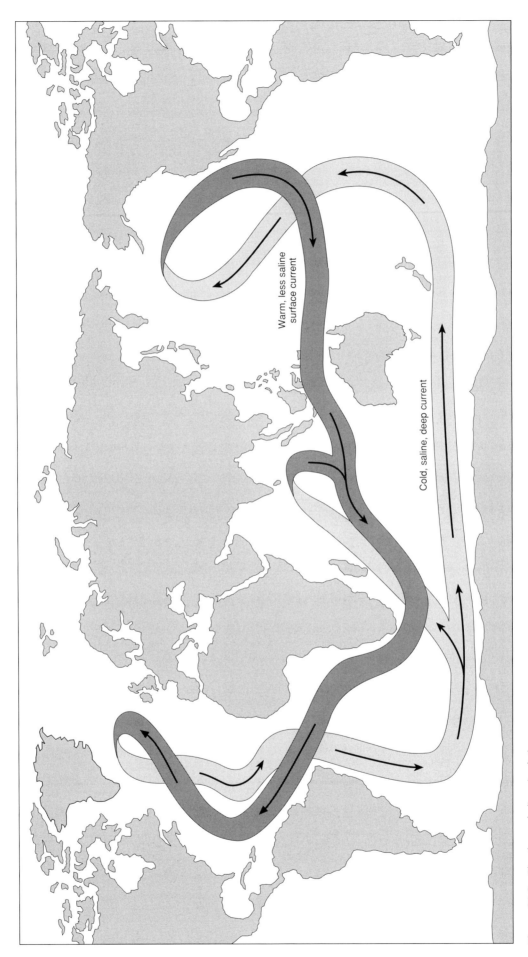

Figure 22.3 The thermohaline circulation.

Topic 22 The marine environment

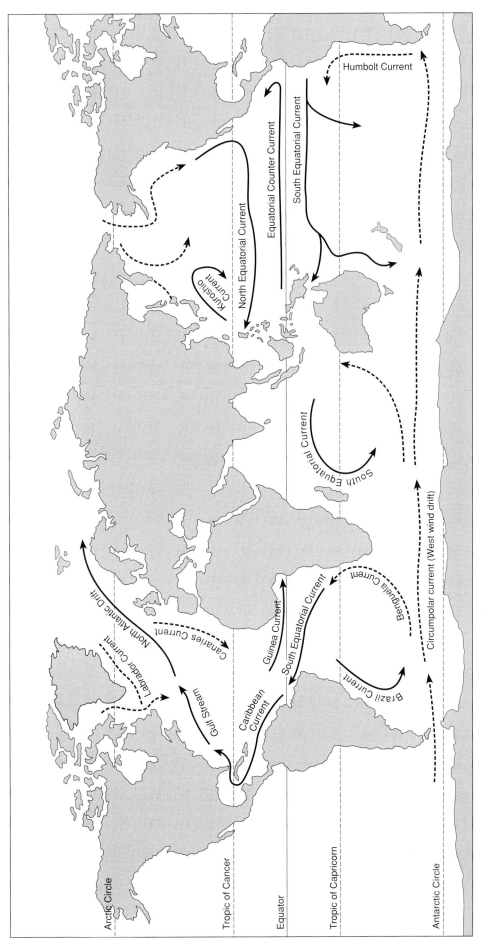

Figure 22.4 The ocean surface currents.

SECTION E

Atmosphere

23. The Atmosphere

The Early Atmosphere

When the Earth first formed the atmosphere was composed of the gases given off by the cooling rock material, mainly nitrogen, water vapour and carbon dioxide with smaller amounts of sulfur dioxide, hydrogen sulfide, methane and ammonia, but crucially no oxygen (Table 23.1). Approximately 3.5 billion years ago the first life evolved and photosynthesis started to produce oxygen as a waste product. The evolved oxygen reacted to oxidise reduced crustal rocks and so did not accumulate in the atmosphere until about 2.5 billion years ago. By 2 billion years ago the atmosphere contained about 3% oxygen, and oxygen continued to accumulate until the present level of 21% was reached. The surface of the early Earth received high levels of UV light which is thought to be critical to the evolution of life, but once oxygen started to accumulate in the atmosphere then ozone also started to form; currently the ozone layer prevents much of the UV light from reaching the Earth's surface.

Table 23.1 Evolution of the Earth's atmosphere.

Billion years ago	Atmosphere
4.5	N_2, CO_2, H_2O (SO_2, H_2S, NH_3, CH_4), but no O_2
3.5	First life and photosynthesis, O_2 evolved
2.5	O_2 starting to accumulate in atmosphere
2.0	3% O_2
Present	21% O_2

Structure of the Atmosphere

The atmosphere is divided into layers based on the temperature profile (Figure 23.1). This is controlled mainly by differences in the adsorption of solar radiation leading to differing degrees of heating.

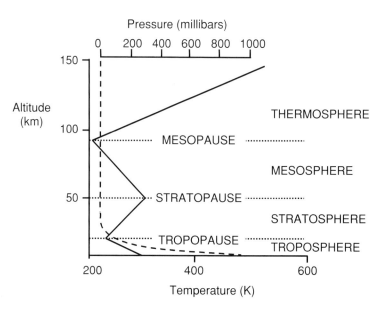

Figure 23.1 Temperature and pressure changes and the structure of the atmosphere.

The lower layer, the *troposphere* is characterised by temperature decreasing with altitude until at the *tropopause* there is a change to an increase in temperature in the stratosphere, due in part to the presence of ozone. The *stratopause* marks the change to falling temperatures in the *mesosphere*, while the outer *thermosphere* is a region of very intense radiation, producing ionised species such as O_2^+, O^+, NO^+, free electrons and high temperatures.

Pressure falls off rapidly with altitude, so 50% of the mass of the atmosphere lies within 5 km of the Earth's surface (Table 23.2).

Table 23.2 Mass profile of the atmosphere.

% of mass	Within km of surface
50	5
90	10
99	50
99.999	80

The Troposphere

The *troposphere* is the part of the atmosphere in which most clouds and weather systems occur (Figure 23.2). It varies in thickness from 9 km at the poles to 16 km at the equator. Supersonic fighters fly close to the tropopause; the Himalayan peaks rise above 9 km and Mexico city is at 2.4 km.

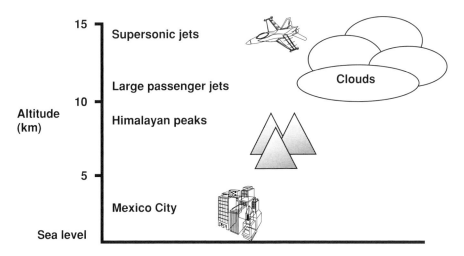

Figure 23.2 The troposphere.

Within the troposphere there is a temperature inversion, the boundary layer, at an altitude of approximately 1 km but varying with meteorological conditions where the falling temperature profile is interrupted by a small increase in temperature (Figure 23.3). This can act to trap pollutant gases close to the surface.

Chemical Composition of the Troposhere

The composition of dry air at sea level is given in Table 23.3. Most of the water vapour is present at lower altitudes as the temperature rapidly falls below freezing point.

Table 23.3 Composition of clean, dry air at sea level.

	% by volume
Nitrogen	78.084
Oxygen	20.946
Argon	0.934
Carbon dioxide	0.321
Neon	0.00182
Helium	0.00052
Krypton	0.00011
Xenon	0.0000087
Methane	0.000125

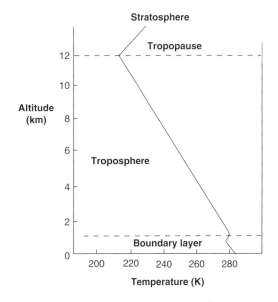

Figure 23.3 The atmospheric boundary layer.

24. Atmospheric Processes

Global Energy Balance

Sunlight striking the Earth's surface at an angle is spread over a greater area than when the sun is directly overhead, reducing the energy input per m^2 in the same way that a shadow is longer. Consequently more of the sun's energy is received at the equator than at the poles (Figure 24.1). This differential warming leads to differences in temperatures and consequently a redistribution of energy towards the poles by means of air currents, evaporation and condensation of water (*condensation cells*), and by ocean currents.

Operation of a Condensation Cell

At the equator, heating of the Earth's surface by the sun causes warm (therefore less dense), moist air to rise. As it does so it cools, causing water to condense, forming clouds and rain. The rising air moves away from the equator, transporting energy towards the poles, continues to cool and therefore sink. The sinking air is compressed as the pressure increases and therefore warms, producing a high pressure area of warm dry air at the Earth's surface. As the air flows back towards the low pressure area created by the rising air it warms and picks up water vapour, so completing the cycle, the overall effect of which is to transport energy polewards (Figure 24.2).

Two tropical condensation cells, known as the *Hadley cells*, provide a good model of the wind patterns and climate within the tropics (Figure 24.3). Rising air at the equator (the *intertropical convergence zone*, ITCZ) produces the high rainfall of the tropics and the zone of sinking air corresponds to the major desert zones at approximately 30°N and 30°S of the equator. A three cell model

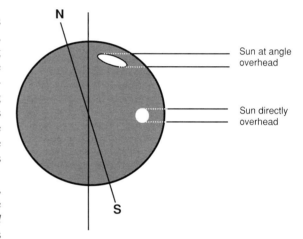

Figure 24.1 Distribution of solar energy at the Earth's surface.

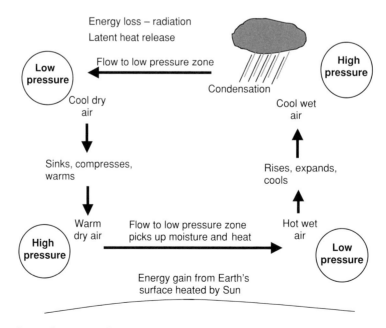

Figure 24.2 Operation of a condensation cell.

Topic 24 Atmospheric processes

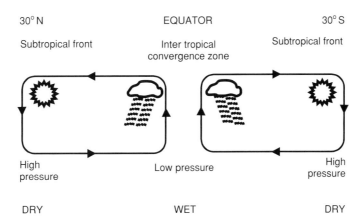

Figure 24.3 The tropical Hadley cells.

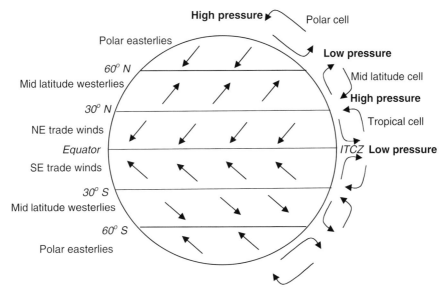

Figure 24.4 The 3 cell model of atmospheric circulation.

is used to explain the major wind patterns at the Earth's surface. In addition to the tropical Hadley cell, driven by hot air rising at the equator, the three cell model includes a polar cell driven by cold air sinking at the pole and a linking ferrel (mid latitude) cell. Because of the *coriolis effect*, the air currents do not move directly north or south. In the northern hemisphere the wind direction is deflected to the right in the southern hemisphere to the left (Figure 24.4).

Dispersal of Atmospheric Pollutants

The dispersal of pollutants in the atmosphere is restricted by the boundaries between the layers of the atmosphere and the boundaries of the condensation cells, particularly at the ITCZ (Table 24.1). The barrier at the ITCZ means that, for example, short-lived pollutants released primarily in the northern hemisphere have lower concentrations in the southern hemisphere, and seasonal changes in carbon dioxide can be clearly seen in the data collected. Pollutants affecting the ozone layer have atmospheric lifetimes of many years which allows them to cross the boundary into the stratosphere.

Table 24.1 Dispersal of pollutants in the atmosphere.

Timescale	Degree of spread of pollutant
Hours	Tens of km within the boundary layer
Days	Thousands of km within the troposphere
Weeks	Troposphere within one hemisphere
Months	Global troposphere, beginning to reach stratosphere

Biosphere

SECTION F

25. Nutrient and Energy Flows in the Biosphere

Living organisms depend on a supply of energy and nutrients to drive their metabolic processes and for the synthesis of new tissue for growth and reproduction. Organisms that can synthesise organic molecules from simple inorganic substrates, carbon dioxide, water and mineral nutrients, are called *autotrophs* and are the *primary producers* in an ecosystem. They can be divided into two groups. The most important are the *photoautotrophs*, which use the energy of sunlight in the process of photosynthesis. In addition there is a small group of *chemoautotrophs* which obtain energy from the oxidation of inorganic chemicals such as sulfide and ammonium. *Heterotrophs* obtain the energy and carbon for metabolism and growth from the breakdown of organic molecules. These are the *consumer organisms* in an ecosystem.

Photoautotrophic Nutrition

These organisms include the plants, the algae and a number of photosynthetic bacteria. The process of photosynthesis utilises the energy of sunlight via the photosynthetic pigment chlorophyll and related molecules to fix carbon dioxide and synthesise organic molecules (Figure 25.1). This complex process can be summarised as

$$6CO_2 + 6H_2O + \text{energy} \rightarrow C_6H_{12}O_6 + 6O_2$$

A range of mineral nutrient elements (see Topic 27) are required to build up the biochemical constituents of the cells.

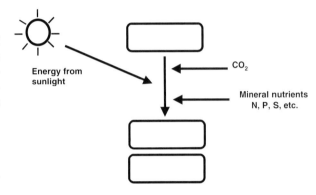

Figure 25.1 Photoautotrophic nutrition.

Chemoautotrophic Nutrition

In chemoautotropic nutrition (Figure 25.2) the energy for carbon dioxide fixation and synthesis of organic molecules comes from the oxidation of a reduced substrate. The sulfur oxidising bacteria, such as *Thiobacillus*, obtain energy from the oxidation of hydrogen sulfide, sulfides and sulfur to sulfate:

$$2S + 3O_2 + 2H_2O \longrightarrow 2SO_4^{2-} + 4H^+ + \text{energy}$$

The nitrifying bacteria *Nitrosomonas* and *Nitrobacter* obtain energy from the oxidation of ammonium and nitrite, respectively:

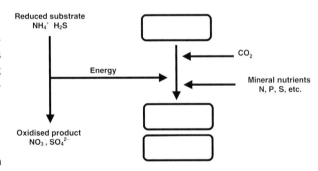

Figure 25.2 Chemoautotrophic nutrition.

$$2NH_4^+ + 3O_2 \longrightarrow 2NO_2^- + 4H^+ + 2H_2O + \text{energy}$$

$$NO_2^- + 0.5O_2 \longrightarrow NO_3^- + \text{energy}$$

The energy obtainable is very small; for example, oxidation of nitrite to nitrate yields just 71 kJ/mol, compared with 2872 kJ/mol for the complete oxidation of glucose. Consequently autotrophic nutrition is only of significance to primary production in the absence of photosynthesis, e.g. round black smokers on the deep ocean bed.

Heterotrophic Nutrition

The animals, fungi and most bacteria obtain their energy from the oxidation of organic molecules (Figure 25.3) in the process of respiration which can be summarised as:

$$C_6H_{12}O_6 + 6O_2 \longrightarrow 6CO_2 + 6H_2O$$

Not all of the organic substrate molecules are oxidised to carbon dioxide and water; some of the molecules are used to synthesise new tissues. Often, significant quantities of heat are generated; for example, composting relies on generation of temperatures of 60–70°C to speed up the decomposition process and kill pathogenic bacteria.

Energy and Nutrient Cycling

The primary producers use the energy of sunlight to form organic molecules using carbon dioxide and taking up a number of mineral nutrient elements from their environment (Figure 25.4). The carbon, the energy in the form of chemical bond energy of the organic molecules and the mineral nutrients are cycled through the ecosystem in a complex food web where the vegetation is consumed by herbivores, which in turn are consumed by carnivores. The leaf litter and dead vegetation, the waste products and eventually the bodies of the animals are broken down by micro-organisms in soils sediments and water. Again, there is a complex food web of prey and predator organisms through which the organic molecules are cycled. Each time the organic molecules are used as a substrate for growth, some energy and carbon are used in the synthesis of new tissues and some carbon is released as carbon dioxide. The decomposer micro-organisms produce the humic material in soils and release the mineral nutrients back into the soil to be utilised by the next generation of plants.

Primary Productivity on Land and in the Oceans

Although the oceans occupy approximately 70% of the Earth's surface area, in contrast to only 30% for the land surface, the standing crop biomass carbon (the carbon present in living organisms) and the primary productivity are both very much lower in the oceans (Table 25.1). Although many factors contribute to this lower primary productivity in the ocean environment, one of the most important is a lack of the essential mineral nutrients.

Primary Productivity in Various Ecosystems

Primary productivity varies enormously in different ecosystems (Figure 25.5). The most productive ecosystems (tropical rainforest, coral reefs and estuaries) have about 20 times the primary productivity of the least productive ecosystems (deserts, open ocean, tundra).

The primary productivity is dependant on a range of factors, one of which will be the factor limiting productivity:

- Light energy received
- Temperature
- Inorganic nutrient supply
- pH
- Gas exchange, CO_2 supply
- Availability of water (in soils)
- Salinity.

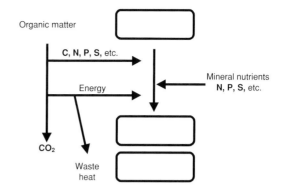

Figure 25.3 Heterotrophic nutrition.

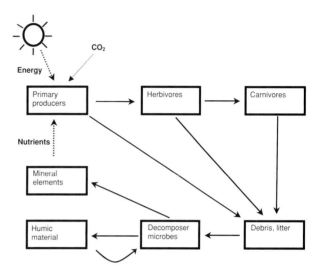

Figure 25.4 Energy and nutrient cycling.

Table 25.1 Primary productivity.

	% of Earth's surface	Biomass (Gt C)	Primary productivity (Gt C/year)
Oceans	70	5	20
Land	30	600	100

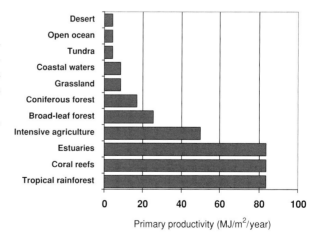

Figure 25.5 Estimates of primary productivity in different ecosystems.

If that limitation is removed (for example, by adding fertilisers) then the productivity increases to a level where some other factor becomes limiting.

26. Diversity of Micro-organisms

Living organisms are divided into six kingdoms:

- Viruses
- Bacteria
- Protozoa
- Fungi
- Plants
- Animals.

All of these interact with the chemistry of the environment; the microscopic organisms and the plants have direct impacts on many of the chemical processes and are vital to the biogeochemical cycling of the elements.

The Diversity of Micro-organisms

Bacteria

This group includes the archebacteria, the eubacteria, the actinomycetes and the cyanobacteria.

Discovered in the 1970s the *archebacteria* are a group of very ancient bacteria restricted to some very extreme environments such as hot springs, black smokers on the deep ocean bed, marshes, the ruminant gut and very saline, very acid or very high temperature environments.

The *eubacteria* (true bacteria), are a very diverse group of single-celled organisms. Typical cells are *coccus* (spherical) or *bacillus* (rod shaped) 0.5–1 µm in diameter (Figure 26.1). They are predominantly heterotrophs and include many disease causing organisms, but are of primary importance in the decomposition of organic matter and the recycling of mineral nutrients. Chemoautotrophic bacteria such as the nitrifying bacteria and sulfur oxidising bacteria are important in the environmental cycles of these elements (see Topic 25).

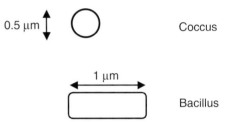

Figure 26.1 Typical bacterial cells.

Bacteria are typically the dominant micro-organisms found in aerobic, neutral pH soil and water environments. There may be as many as 10 to 100 million cells per gram of soil and 1 million cells per mL of water.

Bacteria grow by binary fission (each cell divides to become two) and, with divisions occurring at intervals of a few hours, growth rates can be very rapid. Starting with one cell which divides every hour, after x hours there are 2^x cells (Figure 26.2). Fortunately, numbers are kept in check by predation.

The *actinomycetes* are filamentous bacteria. The filaments (*hyphae*) are typically 0.5–1 µm in diameter and in some species carry single spores or chains of spores. The mass of hyphae is called the *mycelium* (Figure 26.3). These microbes are important decomposers and are responsible for the earthy smell of soil. They are sensitive to mildly acidic conditions and are usually absent from soil more acidic than pH 5.5.

The *cyanobacteria*, also called the *blue-green algae*, are photosynthetic. They may be single cells a few micrometers in diameter or chains of cells forming visible colonies. These photoautotrophs are important primary producers in aquatic environments. Many are able to fix atmospheric nitrogen gas and so are important colonisers in N deficient environments. They may also be responsible for toxic algal blooms in lakes and the oceans.

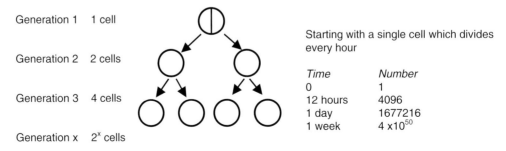

Figure 26.2 Bacterial multiplication by binary fission.

Topic 26 Diversity of micro-organisms

Fungi

The fungi include both unicellar and filamentous forms, ranging from the yeasts to the mushrooms and toadstools. Fugal hyphae are larger than those of the actinomycetes, 5–20 μm diameter, and also produce sporulating bodies containing large numbers of spores. Toadstools are the visible sporing bodies of one group of fungi (the *basidiomycetes*) but there will be many metres of hyphae in the substrate below.

The fungi are heterotrophs and important decomposer microbes. They are frequently tolerant of acid conditions are therefore commonly the dominant microbe in acid environments such as acid soils. The basidiomycetes include the specialised wood rotting fungi and are important in the decomposition of lignin. There may be as much as 100 m of fungal hyphae in a gram of soil.

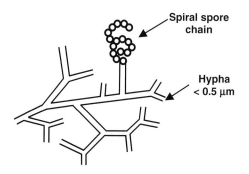

Figure 26.3 Actinomycete mycelium consisting of hyphae and spores.

Lichens

Lichens are a symbiosis of a fungal and a cyano bacterium partner. They are photosynthetic, and many can fix atmospheric nitrogen so are important colonisers of bare rock and branches of trees. They are sensitive to atmospheric pollution and are used as biological indicators of atmospheric pollution levels.

Mycorrhiza

Mycorrhiza are a symbiosis of fungi with plant roots. Most plant roots are enveloped in a sheath of fungal tissue with some of the hyphae penetrating the root between the cortical cells (*ectomycorrhiza*) while in other cases the fungal hyphae enter the root cells (*endomycorrhiza*). The mycorrhizal symbiosis aids the plant root in the uptake of nutrients, particularly phosphate, from soil.

Protozoa

The protozoa are a mixed group of single celled (up to 100 μm diameter) organisms that have at various times been classified as plants and animals. They include both photoautotrophs and heterotrophs (decomposers and predatory organisms) that contribute to the phytoplankton and zooplankton, and are common in soils and sediments (Figure 26.4).

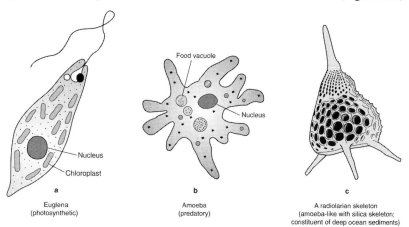

Figure 26.4 Some examples of Protozoa.

Definitions

Micro-organisms	Microscopic organisms (microbes) – bacteria, fungi and protozoa
Flora	Plants
Fauna	Animals
Mesofauna	In soils those animals that are small enough to move through the existing pores system
Macrofauna	In soils those animals which create their own tunnels by burrowing and therefore modify soil structure, e.g. earthworms
Phytoplankton	Microscopic, free floating or motile but go with the water currents, photosynthetic (autotrophs) – cyanophytes, protozoa and plants
Zooplankton	Microscopic or larger, free floating or motile but go with currents, consumers (hetertrophs) – protozoa and animals, e.g. insect larvae, crustaceans
Macrophyte	Larger aquatic plants, rooted or floating
Nekton	Actively swimming animals

27. Inorganic Plant Nutrients

Plant Nutrients

The inorganic elements required by plants are subdivided into the *essential nutrients* (those without which the plant cannot develop and reproduce normally) and the *beneficial nutrients* (those which improve growth of the plant but are not essential, and those which are an essential requirement of animals consuming the plants but not to the plants themselves). They are also subdivided into *macronutrients* (present in the plant at concentrations of greater than 1000 mg/kg) and *micronutrients* or *trace elements* (less than 100 mg/kg). A total of 16 elements are considered essential for higher plants: C, H, O, N, P, S, Ca, Mg, K and Cl are essential macronutrients and the essential micronutrients are B, Cu, Fe, Mn, Mo and Zn. Sodium is beneficial for some plants (sugar beet) and is essential to animals. Silicon is beneficial to many grasses and is essential to diatoms. Aluminium is beneficial to many ferns. Cobalt is an essential requirement for nitrogen-fixing micro-organisms and is therefore beneficial to legume plants. Chromium, selenium and iodine are required by animals but not by plants.

Uptake of Nutrients

Carbon, hydrogen and oxygen are supplied as carbon dioxide and water, while the remainder are present in soil or water as dissolved ions. The uptake of nutrients from the soil is a complex mixture of chemistry (chemical forms and reactions of the nutrient ions), biology (development of the root system and uptake by roots) and physics (the movement of ions through the soil towards the root surface).

Nutrient Availability

Roots and micro-organisms take up nutrient ions from the surrounding solution. The solution concentration of most of these ions is buffered by a supply of ions held on exchange sites and adsorbing surfaces in the soil (see Topics 7 and 8). This represents the pool of plant available nutrients. Non-available forms of nutrients, such as those in the matrix of minerals, clays and oxides or the soil organic matter, are slowly released into the plant available pool by weathering and microbial breakdown (Figure 27.1).

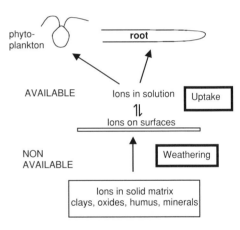

Figure 27.1 The chemical availability of plant nutrients.

Root Development

The development of an extensive root system is critical to a plant's ability to fully exploit the pool of chemically available nutrients present in a soil profile. Only those nutrients sufficiently close to a root will be taken up, even if chemically available. One of the major factors determining the form and extent of the root system is the type of plant. Most plants produce greater numbers of roots in the upper part of the soil profile; however, *monocotyledonous* plants (e.g. grasses) produce large numbers of adventitious roots from the stem bases, leading to a very dense surface root system. By contrast *dicotyledonous* plants (broad-leaf plants) produce a main tap root with extensive branching, which leads to less dense surface rooting (Figure 27.2).

Adverse soil conditions can also reduce the ability of roots to exploit the available nutrient pool. In surface horizons, adverse pH, dryness and salinity may influence root development. In subsurface horizons, poor structure and the presence of shallow pans, rock and water table can also be important. Root systems will avoid areas that are anaerobic or where toxic pollutants are present, but a developing root system will continuously grow into new areas of soil (Figure 27.3).

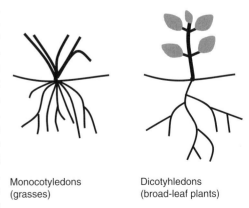

Figure 27.2 The form of plant root systems.

Topic 27 Inorganic plant nutrients

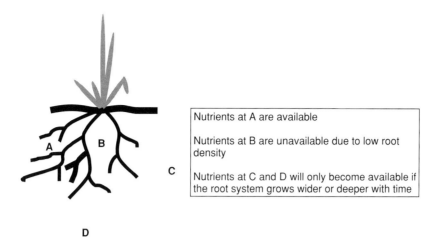

Figure 27.3 Spatial variability in nutrient availability.

Nutrient Transport to Roots

Nutrients move to the root surface by *mass flow* and *diffusion*. Plants continually lose water from their leaves (*transpiration*) and therefore take up water through the root system. This flow of water brings with it the ions dissolved in the soil solution. The contribution of mass flow can be calculated as volume of water transpired × concentration in soil solution. In a well-fertilised soil this will typically provide Ca and Na in excess of plant requirements, NO_3^- and Mg about equal to requirements, but K at 10% and P only 1% of requirements.

Uptake of nutrients at the root surface generates a low solution concentration, which in turn leads to diffusion of ions from areas of high concentration in the soil towards the root. Relative to the diffusion in water the diffusion of ions in a soil pore system is reduced by the water filled porosity (Φ), the tortuosity of the pathway (f) and the ratio of concentration of diffusible ion in solution to the ion on exchange sites:

$$D_{\text{soil}} = D_{\text{water}} \, \Phi f \, \frac{[\text{Ion}] \text{ in solution}}{[\text{Ion}] \text{ on exchange-sites}}$$

Only the smaller water-filled pores will transmit ions in solution and the complexity of the pathway of continuous water–filled pores is expressed in the tortuosity term (Figure 27.4).

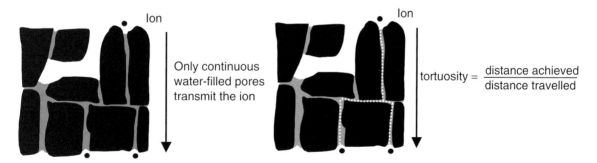

Figure 27.4 Effects of porosity and tortuosity on diffusion pathway.

These two factors influence the diffusion of all ions in the same way. The major differences in soil diffusion coefficients arise from the ratio of the concentration of the ion in solution to those held on exchange sites. In the case of nitrate all the ions are in solution so diffusion is relatively fast, while for phosphate little is in solution so diffusion is very slow. Potassium is intermediate in behaviour (Table 27.1).

The very slow diffusion of phosphate in soils means that it can only travel a few millimetres through the soil to a root and thus diffusion is unable to supply sufficient P for the plant's needs. Root hairs (projections of the epidermal cells on the root surface) greatly increase the volume of soil within a few millimetres of an absorbing surface, thereby increasing the P supply. Even so, many plants are unable to take up sufficient P and are dependent on mycorrhizal fungi to obtain sufficient P. These symbiotic fungi grow in and around the root tissues and surrounding soil and are able to transport P back to the root.

Table 27.1 Diffusion coefficients in water and soil (typical values).

Ion	D_{water} (cm²/s)	D_{soil} (cm²/s)
NO_3^-	2×10^{-5}	2×10^{-6}
K^+	2×10^{-5}	2×10^{-7}
$H_2PO_4^-$	1×10^{-5}	1×10^{-9}

Chemical, Physical and Biological Interaction

SECTION G

28. Temperature of Environmental Systems

Temperature affects the rate of chemical and biochemical reactions, and therefore the rate of most processes in the environment. Seed germination, seedling growth, plant root and shoot development and the uptake of nutrients and water, microbial processes such as organic matter decomposition, and the mineralisation of inorganic nutrients and peat accumulation are all affected by temperature. In general, the reaction rate approximately doubles for each 10°C rise in temperature up to an optimum temperature above which the rate falls with further increase in temperature. Chemical processes such as weathering and the solubility of salts and gases in water are also temperature dependent.

Surface Energy Balance

The temperature of an environment depends on the intensity of solar radiation arriving at the Earth's surface and transfers of energy between the surface and the atmosphere. Incoming solar energy will vary with the time of day, the season, the latitude and the aspect (as these influence the angle at which sunlight strikes the Earth's surface) and importantly the cloud cover. The proportion of the energy absorbed by the surface depends on its albedo (reflectivity) (Table 28.1).

Table 28.1 Typical albedo values.

Bare soil	0.2–0.5
Vegetation	0.15–0.3
Snow (fresh)	0.9
Water surface	0.03–0.3
Earth's average	0.35

All bodies emit light energy at temperatures above 0 K; the sun at 6000 K emits predominantly visible light, while the Earth at 290 K emits long wavelength infrared radiation back out into space. This back radiation is absorbed by atmospheric carbon dioxide and water vapour (the greenhouse effect – see Topic 49). Further transfers of energy between the Earth's surface and the atmosphere occur by conduction and through the evaporation and condensation of water (latent heat of vaporisation of water is 41 kJ/mol) (Figure 28.1).

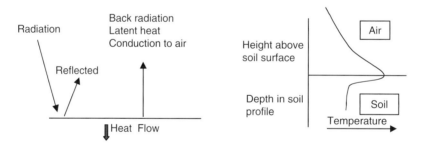

Figure 28.1 Daytime energy balance and temperature profile at a bare soil surface.

During the daytime (Figure 28.1), the soil surface is the hottest part of the system, heating the air above and the soil profile below, while at night the opposite is true (Figure 28.2). As a result of energy losses by back radiation, the soil surface becomes the coldest part of the system. In water bodies the effects are similar, but the sun's energy can be absorbed throughout the depth to which light penetrates.

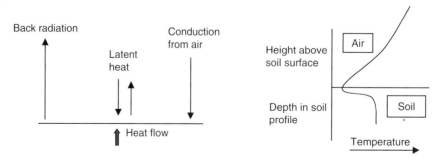

Figure 28.2 Night-time energy balance and temperature profile at a bare soil surface.

Heat Flow Through a Soil Profile

The transfer of energy through the soil profile depends on the thermal conductivity of the soil, but the change in temperature depends on the specific heat. Although silica is a good thermal conductor, a pile of sand is a poor conductor because of the poor grain-to-grain contact. In a soil, the thermal properties of the solid matrix (sand, silt and clay) are modified by the presence of air or water in the pores. Air is a good insulator, while water is a good thermal conductor but has a high specific heat. Thus a dry sandy soil will be a poor conductor with a low specific heat, resulting in a large daily variation in temperature at the surface and a steep change in temperature down through the profile, whereas a wet clay soil will be a good conductor with high specific heat, leading to a smaller daily variation in surface temperature and a less steep change with depth.

In a soil it is the surface which both receives and loses energy, so it responds directly to changes in incoming solar radiation; on a clear day maximum surface temperature will occur at midday. Maximum soil profile temperatures occur progressively later in the day and are progressively damped with depth. At 30 cm no daily temperature cycle is seen, but seasonal variations can be observed (Figure 28.3).

Air temperature is measured at 1.5 m above soil level and soil temperature at 30 cm depth. The daily and vertical variations in soil temperature have important implications for the effects of temperature on both chemical and biological processes in soils.

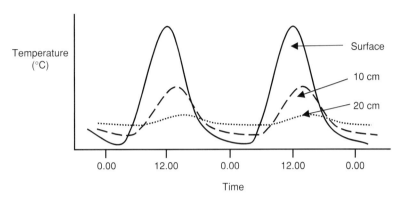

Figure 28.3 Daily temperature cycles in a bare soil.

Thermal Stratification of Lakes

Warm water is less dense than cold water, so when a deep calm lake warms in early summer a layer of warm less dense water forms (the *epilimnion*) and does not mix with the underlying colder water (the *hypolimnion*). In autumn, as solar energy inputs decrease and the surface water cools, thermal stratification breaks down, forming a well mixed water column (Figure 28.4).

Water has a maximum density at 4°C and ice is less dense than water; consequently in winter ice forms at the surface while the water at depth remains above freezing at 4°C. A *monomictic* lake displays one period of free mixing per year and a *dimictic* lake two.

In summer the *thermocline* prevents mixing of gases and dissolved salts between the epilimnion and the hypolimnion. Impeded downward movement of oxygen may lead to the development of anaerobic conditions in the hypolimnion and sediments, while phytoplankton in the hypolimnion are isolated from the supply of inorganic nutrients provided by decomposition processes in the sediments.

In the oceans, a permanent thermocline exists at latitudes less than about 60° N and 60° S, which is slightly shallower close to the equator. In summer a seasonal thermocline often develops above the permanent thermocline (Figure 28.5).

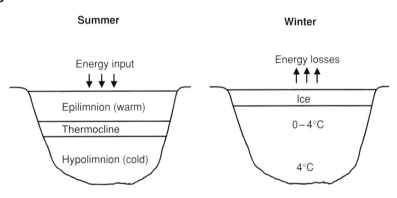

Figure 28.4 Thermal stratification of a lake system.

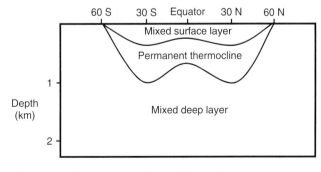

Figure 28.5 The permanent ocean thermocline.

29. Soil Water

The Soil–Plant–Water System

Figure 29.1 represents the hydrological cycle in the soil–plant system. Light precipitation falling on the land surface may be intercepted by, and evaporate from, the vegetation cover, and so not reach the soil below. More intense precipitation will reach the soil surface. The steeper the slope and the more intense the rainfall, the more likely the water is to run off the surface downslope without infiltrating into the soil profile. Water that infiltrates into the soil profile may be stored within the soil pore system, flow downslope within the soil profile or drain through the soil and rock strata to the water table. Water running off the soil surface or draining through the soil forms streams, rivers and lakes, and flows eventually to the oceans. Plants take up water from the soil profile and evapotranspiration from land and evaporation from water surfaces complete the cycle.

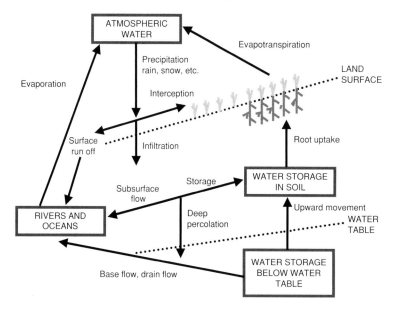

Figure 29.1 Cycling of water in the soil–plant system.

From the perspective of the plant, the main factors influencing growth are:

- the total annual rainfall and the seasonal pattern
- the position of the water table (at depth or within the soil profile)
- the ability of the soil to drain excess rainfall during wet periods
- the ability of the soil to store plant available water in dry periods

Water in Soil Pores

The total pore space of a well structured soil is typically 40–50% of the volume. The pores within and between soil peds may contain water or air, so the more water that is held the less space there is available for air. Water tends to occupy the finer pores, while the larger pores are air filled.

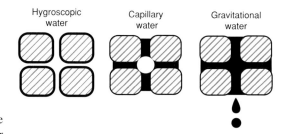

Figure 29.2 Mechanisms holding water in the soil pore system.

Hygroscopic water is chemically bonded to the surfaces of soil particles such as clays and oxides (Figure 29.2). It is only a few percent of the total water in the soil, but is too strongly bound to be plant available and is lost only when soil is heated in a furnace at 400–500°C. The *capillary water* is held by capillary attraction within the soil pores. The smaller the pore the more strongly the water is held within the pore; thus water in the larger pores is plant available but that held in the smaller pores is too strongly held. *Gravitational water* is free draining water that will drain through the soil profile under gravity. At saturation all the pores contain water. If the gravitational water is allowed to drain, after a few days a field soil will reach a steady water content – *field capacity*. Plants can then extract capillary water from the pore system until the remaining water is too strongly held and the plant can extract no more water and so wilts and dies. This soil water state is termed *permanent wilting point*.

The *available water capacity* of a soil profile is the water held in the profile between field capacity and permanent wilting point down to the limit of rooting depth of the plant. It represents the potential of the soil to store plant available water during periods of low or no precipitation

The *soil moisture deficit* is the water (rainfall or irrigation) required to bring the soil profile back to field capacity.

Soil Water Potential

The *soil water potential* (ψ) is a measure of how strongly water is held in the soil pore system. Strictly it is a thermodynamic property, the change in free energy per mole from water in the standard state to the soil water, but it is more commonly measured and interpreted as the negative pressure or suction that is required to extract water from the soil pores. The values are expressed in units of pressure (bar) and are negative; a large negative value represents a large suction required to extract water strongly held in the small pores of the pore system. At saturation, water occupies all the pores and the soil water potential is 0 bar. Field capacity corresponds to -0.05 bar and permanent wilting point to -15 bar.

As the primary mechanism holding water in the soil over the range from field capacity to permanent wilting point is the capillary attraction of the soil water for the soil pore system, the soil water potential can be related to pore diameter through the following equation:

$$d = \frac{-4T}{\psi}$$

where T is the surface tension of water and d is the pore diameter. Pores of diameter larger than d will be air filled, while pores of diameter less than d will be water filled. This means that as wet soil dries, soil water is removed from progressively smaller pores where it is more strongly held and the soil water potential becomes a progressively larger negative value. As a dry soil wets, water first enters the smallest pores then fills progressively larger pores. The soil water potential becomes a progressively smaller negative value until it reaches zero at saturation.

The storage of plant available water in a soil profile depends on the presence of medium sized pores that hold water between the soil water potentials corresponding to field capacity and permanent wilting point.

Water Movement

Water moves through the soil pore system in response to differences in soil water potential. The difference may be generated by:

(1) Differences in soil water content (the *matric potential*) – water will flow from moist soil to dryer soil
(2) Differences in dissolved salts concentration (the *osmotic potential*) – water will flow from regions of low dissolved salts concentration to regions of high dissolved salts concentration
(3) Gravity (the *gravitational potential*) – water will drain downwards through the soil profile.

The total potential is the sum of these three components and water movement depends on the difference in total potential. For example, water may move upwards through the soil where a dry or high salt containing surface layer generates an upward soil water potential gradient that is greater than the downward gravitational potential.

The volume of water flowing through a capillary pore depends on the fourth power of the radius of the pore. For example, if two pores differ by a factor of 10 in their radii the larger pore will transmit 10 000 times more water than the smaller pore. Consequently the drainage of water is highly dependant on the presence of a system of continuous large pores and on the soil water content, as these pores will only be water filled under wet conditions.

A well structured (see Topic 17) or coarse textured (see Topic 16) soil will have the necessary large diameter pores to have good drainage, whereas a poorly structured or fine textured soil will have poor drainage. All soils will drain best when close to saturation, as all the largest pores will be water filled and contribute to water flow. As the soil dries the large pores become air filled and no longer contribute to water flow so the rate of water flow falls very quickly (Figure 29.3).

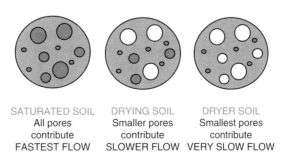

Figure 29.3 Cross section through a soil core showing the effect of pore size on drainage.

30. Aeration and Gas Exchange

The Soil Atmosphere

Respiration by micro-organisms, animals and plant roots within a soil profile leads to a decrease in oxygen concentration and increase in carbon dioxide concentration of the air in the soil pore system compared to the atmosphere (Table 30.1). A continual exchange of gases between soil and atmosphere is required to maintain aerobic conditions in the soil. The lower oxygen concentration in pasture soil is due to a higher respiration rate and oxygen demand by greater numbers of micro-organisms and plant roots.

The air filled porosity of a soil depends on its texture, structure and moisture content, but values of 10 to 40% are typical (Table 30.2). Comparing these values with a typical respiration rate (10–20 L oxygen/m²/day for a vegetated soil in summer) shows that the soil pore system may contain sufficient oxygen to sustain aerobic respiration for approximately 1 week and highlights the importance of the replenishment of oxygen from the atmosphere.

Table 30.1 Composition of the soil atmosphere compared with the free atmosphere.

	O_2 (%)	CO_2 (%)
Free atmosphere	21	0.035
Cultivated soil	20	0.6
Pasture soil	18–20	0.5–1.5

Table 30.2 Typical air filled porosity of soils.

Soil	Air filled porosity at field capacity (%)
Clay	10
Sandy loam	30

Gas Exchange in Soil

The primary mechanism of gas exchange in a soil profile is by diffusion of gases through the air filled pore space. Mass flow, driven by small pressure changes created by the turbulence of wind blowing over the soil surface and the movement of water containing dissolved gases, is of minor importance.

Relative to the diffusion in air, the diffusion of gases in a soil pore system is reduced by the air filled porosity (Φ) and the tortuosity of the pathway (f)

$$D_{\text{soil}} = D_{\text{air}} \Phi f$$

Table 30.3 Diffusion coefficients of oxygen and carbon dioxide in air and water at 25°C.

	Diffusion coefficient (cm²/s)	
	O_2	CO_2
Air	0.21	0.16
Water	2.3×10^{-5}	1.7×10^{-5}

The diffusion coefficient for oxygen in air is 10 000 times the diffusion coefficient in water (Table 30.3); hence only the air filled pores contribute to gaseous diffusion, and the complexity and continuity of the air filled pore system is expressed in the tortuosity term (Figure 30.1).

Soil moisture content has a major effect on the air filled pore system of a soil and thus on the rate of gaseous diffusion (Figure 30.2).

When a dry soil wets, the smallest pores fill with water. Water filled necks

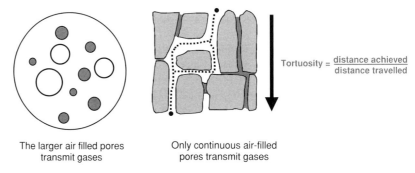

Figure 30.1 Pathway for diffusion of gases through soil.

in pores will effectively prevent that pore from transmitting gases, so the diffusion rate decreases. As the soil wets further, the diffusion rate continues to decrease until the soil is nearly saturated. While there is one remaining air filled pathway, gas movement through that pore is so much faster than through all the water filled pores that the air filled pathway controls the diffusion rate. Once the soil is fully saturated then diffusion must take place through water:

$$D_{\text{soil}} = D_{\text{water}} \Phi_{\text{total}} f$$

All the pore space contributes and the pathway is at its least complex, but the diffusion coefficients of the common gases in water are approximately 10 000 times slower than through air, so gaseous diffusion through a saturated soil is extremely slow. The slow diffusion of oxygen into a water saturated ped, combined with the uptake of oxygen by the soil micro-organisms, means that only the outer part of the ped remains aerobic. Peds with a diameter greater than about 1 cm are likely to have an anaerobic centre when the soil profile is at field capacity, with the macropores between the peds air filled and the micropores within the ped itself water saturated.

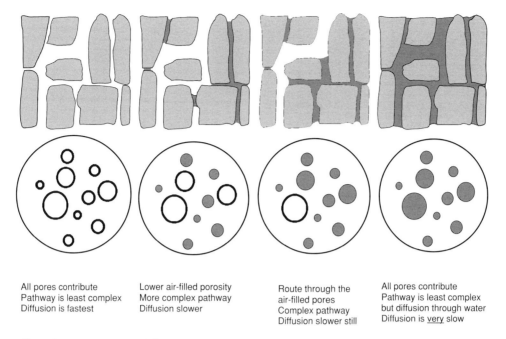

All pores contribute
Pathway is least complex
Diffusion is fastest

Lower air-filled porosity
More complex pathway
Diffusion slower

Route through the
air-filled pores
Complex pathway
Diffusion slower still

All pores contribute
Pathway is least complex
but diffusion through water
Diffusion is very slow

Figure 30.2 Effect of water content on diffusion through soil.

Aeration of Water Bodies

Oxygen has low solubility in water (9.07 mg/L at 20°C in equilibrium with air at 1 bar), so dissolution and vertical movement of atmospheric oxygen is important in maintaining aerobic conditions in the water column and underlying sediments. Carbon dioxide is required for photosynthesis by the phytoplankton, so its concentration is also important. Since gas diffusion through water is so slow, gas movement in water bodies is very dependant on mixing of the water. At the surface oxygen is entrapped by wave action; and vertical mixing of the water carries oxygen into the water profile. The surface waters of the oceans tend to be supersaturated due to the vigorous wave action; so too are fast flowing streams running over rocky bottoms, while slow moving deep rivers and lakes may be less well aerated at depth. In static water the balance between photosynthesis and respiration may be important in determining oxygen and carbon dioxide concentrations.

Thermal stratification of lakes has important effects on gas concentrations in the water and sediments (see Topic 28). Although the epilimnion has good gas exchange with the atmosphere, providing oxygen for respiration and carbon dioxide for photosynthesis, the thermocline blocks gas exchange with the hypolimnion, which may result in the development of anaerobic conditions in the hypolimnion and sediment surface (Figure 30.3).

Even where the water column is well areated and well mixed, only a thin surface layer of the sediments will be aerobic (Figure 30.4). The slow diffusion of oxygen into the water saturated pore system of the sediment, combined with the high oxygen demand of the micro-organisms breaking down organic matter in the sediments, means that the oxygen diffuses only a few millimetres into the surface of the sediment before being consumed, and below this layer the sediment is anaerobic. Products of anaerobic metabolism such as ammonium compounds, organic acids and hydrogen sulfide diffuse up through the sediments and into the water column. These compounds can be toxic to aquatic organisms, but fortunately many are oxidised in the aerobic surface layer of the sediments, thus reducing their impact.

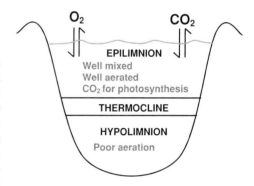

Figure 30.3 Aeration in a stratified lake.

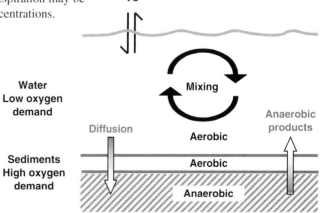

Figure 30.4 Aeration of sediments.

Environmental Cycles

31. The Carbon Cycle

The dominant component of the carbon cycle is the biosphere. Carbon cycles through it to and from the atmosphere and lithosphere as decomposition and waste products. Figure 31.1 shows the principal components of the environmental carbon cycle and the major fluxes between them. Carbon in the lithosphere plays very little part in the overall cycle in the short term. The shift of carbon from the lithosphere to the surface environment is seen by the enrichment factor of 50 between the average crustal content of 0.05% and the average soil content of 2.5%.

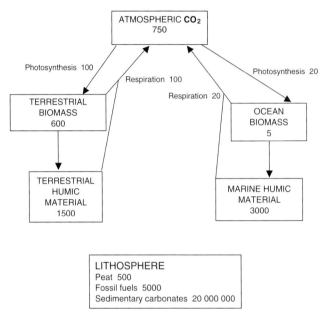

Figure 31.1 The principal reservoirs (in boxes) and fluxes in the carbon cycle (reservoir units gigatonne; flux units Gt/year).

Photosynthesis and Respiration

Photosynthesis is the utilisation of atmospheric carbon dioxide by plants and some autotrophic micro-organisms, or primary producers, which converts the carbon into carbohydrate (see Topic 25):

$$CO_2 + H_2O + \text{energy} \longrightarrow \text{Carbohydrate} + O_2$$

The early atmosphere on earth contained large amounts of CO_2 but no O_2 (see Topic 23). Photosynthetic organisms brought about the shift of carbon from the atmosphere to the biosphere and lithosphere, and released oxygen into the atmosphere.

Respiration is the reverse process, whereby heterotrophic organisms use the carbohydrate to produce energy and carbon dioxide as a waste product:

$$\text{Carbohydrate} + O_2 \longrightarrow CO_2 + H_2O + \text{energy}$$

Under current environmental conditions, these two processes are roughly in balance, with approximately 120 gigatonne (Gt) of carbon moving in each direction per year. (Note that the current concern over fossil fuel burning adding CO_2 into the atmosphere represents a relatively small flux of carbon compared to this (about 5 Gt per annum), but it is sufficient to imbalance the natural steady state – see Topic 49.)

Topic 31 The carbon cycle

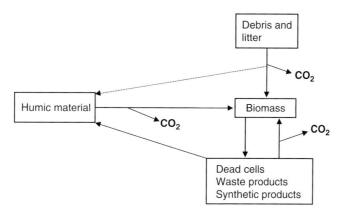

Figure 31.2 Decomposition processes.

Decomposition Processes

Waste and decay products from the biosphere undergo decomposition by heterotrophic organisms (see Topic 25), which use the carbon as an energy source for respiration (Figure 31.2). The products of this process are mainly carbon dioxide and humic material, with some new biomass also being produced.

The important groups of heterotrophs, or consumers, involved in decomposition processes are:

- *Animals* – worms, insects, zooplankton
- *Mico-organisms* – bacteria, fungi, actinomycetes
- *Predators* – amoebae, zooplankton.

They degrade organic material in order to gain energy, releasing CO_2 and inorganic forms of N, P and S (see Topics 32, 33, 34). The organic substrate that is decomposed (mainly debris, litter and dead cells) consists of a range of compounds of widely differing stabilities. Simple carbohydrates such as sugars, storage carbohydrates such as starch and glycogen, and proteins are major components of plant and animal tissue, and therefore of this fraction; they are readily broken down by common enzyme systems. Structural carbohydrates, such as cellulose, hemicellulose and lignin, are more complex molecules and therefore more stable, as the enzymes required to degrade them are less common. For example, fungi are generally better than bacteria at degrading lignin, and so decomposition of wood is initially carried out predominantly by certain groups of fungi. Two broad groups of heterotrophic organisms can be recognised, based on their ability to attack organic molecules and their ability to respond to inputs of organic material. The *autochthonous population* is stable and has a slow rate of growth. These organisms tend to attack the more resistant substrates, causing the slow, long term decomposition of organic matter in the environment. The *zymogenous population* is variable and has a fast rate of growth. It can lie dormant for long periods, as it forms resistant spores which can be activated in response to inputs of organic matter. These organisms tend to degrade simple substrates.

Inorganic Components of the Carbon Cycle

In addition to the organic carbon cycle described above, there are components based on the inorganic chemistry of carbon – carbonates and carbon dioxide – and these are particularly important in the oceans.

Carbon dioxide dissolves slightly in water to form carbonic acid (H_2CO_3), which ionises to give bicarbonate (HCO_3^-) and carbonate (CO_3^{2-}) ions (see Topic 11):

$$H_2CO_3 \rightleftharpoons HCO_3^- + H^+ \rightleftharpoons CO_3^{2-} + 2H^+$$

Calcium ions react with carbonate to produce calcium carbonate, which is a highly insoluble salt and is used by many marine organisms to form shells. When these organisms die, the shells persist and sediment out to form calcium carbonate deposits, which over geological time are converted to chalk and limestone rocks. As can be seen from Figure 31.1, a vast amount of carbon has been fixed into this form, which is now by far the highest reservoir of carbon in the environment. Because of its insolubility, and because much of the rock does not interact with the surface environment, this carbon is essentially inert.

32. The Nitrogen Cycle

As with carbon, the biosphere and organic matter play a central role in the nitrogen cycle (Figure 32.1). The atmosphere is the major reservoir of N (Table 32.1), but most of that is inert and takes no part in N cycling. There are important inorganic transformations, often microbially driven, in the soil and water environments. Although N in the crustal rocks is a major reservoir, it contributes little to the N cycle, other than by release of ammonia by volcanic action. The long term shift of N to the surface environment is seen by the enrichment factor of 80 between the average crustal content of 0.0025% and the average soil content of 0.20%.

Table 32.1 Reservoirs of nitrogen in the environment.

Environmental reservoir	N content (Gt)	% of total N
Atmosphere	3.9×10^6	72
Crustal rocks	1.5×10^6	28
Hydrosphere	2.3×10^4	0.4
Biosphere	280	0.005
Soils	240	0.004

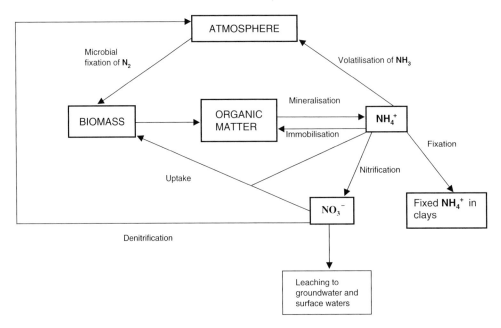

Figure 32.1 The nitrogen cycle.

Mineralisation and Immobilisation of Nitrogen

Nitrogen in biomass tissue is predominantly amino N (NH_2—N) in proteins and amino acids, with smaller amounts in molecules such as nucleic acids, chorophyll, adenosine di- and tri-phosphates. As this tissue decomposes, nitrogen is released as ammonia (NH_3), which is rapidly hydrolysed to the ammonium ion (NH_4^+) under most environmental conditions. This process is driven by micro-organisms, which use the C and N released to form new tissue. *Mineralisation* occurs when there is excess N released over microbial requirements, whereas *immobilisation* occurs if all the N is utilised by the micro-organisms. The balance between mineralisation and immobilisation depends on the C:N ratio. In fresh biomass tissue this tends to be wide, typically C:N > 25:1, and the process tends towards *immobilisation* – i.e. there is relatively a lot of C for use in tissue production and consequently there is a high demand for N by the micro-organisms. If the biomass tissue is high in N (e.g. tissues of leguminous plants, which can fix atmospheric N_2 – see below) or if the C:N ratio is narrow (e.g. humic material, typically C:N < 10:1), then the amount of N released from decomposing organic material is greater than microbial requirements, resulting in *mineralisation*.

Nitrification and Denitrification

In aerobic environments above pH 5.5 the ammonium released by mineralisation is oxidised to nitrite and then nitrate by two groups of chemoautotrophic bacteria. These organisms use CO_2 as their carbon source and derive their energy from the nitrification process. The first step is carried out by bacteria of the genus *Nitrosomonas*:

$$2NH_4^+ + 3O_2 \rightarrow 2NO_2^- + 4H^+ + 2H_2O + \text{energy}$$

and the second step to nitrate by the genus *Nitrobacter*:

$$2NO_2^- + O_2 \rightarrow 2NO_3^- + \text{energy}$$

The second reaction proceeds faster than the first, so nitrite (NO_2^-) does not normally accumulate in environmental systems.

Nitrate is a soluble ion, which is highly mobile in the environment and so readily leached into the hydrosphere. This can cause problems in both drinking water (see Topic 40) and surface waters (see Topic 42). Under moderately reducing conditions (see Topic 12), it is converted to N_2, or oxides of nitrogen, principally nitrous oxide (N_2O) and nitric oxide (NO). These are gases and so this is a route for N to be transferred from the terrestrial and aquatic environments into the atmosphere. Typical rates of loss of N by denitrification in agricultural soils are of the order of 10–30 kg N/ha/year. Nitrous oxide is a potent greenhouse gas (see Topic 49).

Volatilisation of Ammonia

Under alkaline conditions the NH_4^+ ion is converted to ammonia gas, which can be lost to the atmosphere by volatilisation.

$$NH_4^+ + OH^- \rightleftharpoons NH_3 + H_2O$$
$$pKa = 9.25$$

Ammonia can be lost when organic matter high in N breaks down, and in soils when urea fertiliser, manure or sewage sludge is added. Although NH_3 itself is alkaline, it reacts with acid gases in the atmosphere and is converted back to the ammonium form. When this returns to soil and surface water it is rapidly nitrified, which is an acidifying process. Atmospheric NH_3 therefore contributes to the effect of acid rain (see Topic 52).

Microbial Nitrogen Fixation

The N_2 molecule is very stable and considerable energy is required to break the N≡N triple bond. This is done industrially by use of high temperature, high pressure and a catalyst, but some micro-organisms have an enzyme that can do this under normal environmental conditions. Some of these micro-organisms are free-living, for example the aerobic bacteria *Azotobacter*, the facultative anaerobe *Klebsiella*, the obligate anaerobe *Clostridium* and the photoautotrophic cyanobacterium *Anabaena*; they are commonly found in soils and waters. Other N fixing organisms exist in association with plants. Of particular importance are root nodule organisms such as *Frankia*, associated with alder and bog myrtle, and *Rhizobia*, associated with legumes such as clover, peas, beans and lupin. It is estimated that on a global basis nitrogen fixing organisms contribute 0.15 gigatonne of N per year, or 60% of the N inputs to the biosphere.

Fixation of Ammonium by Clay Minerals

The hydrated ammonium cation in solution is similar in size to the hydrated potassium cation (0.53–0.54 nm), so can readily be fixed in the interlayer space of clay minerals (see Topics 3, 7 and 35). This is an immobilising process for N in the environment, with the NH_4^+ ion being released only when the clay minerals weather and other ions can enter the interlayer space and ion exchange can occur.

Bioavailability of Nitrogen

Organisms other than the N-fixing species obtain their nitrogen as NO_3^- or NH_4^+ ions in solution. Concentrations of these ions are usually low in natural waters unless there has been leaching of nitrate. Nitrate is the main anion in soil solution in agricultural soils. Bioavailability of N is dependent on mineralisation from organic matter releasing a slow but regular supply that can be used by plants and micro-organisms. It is therefore difficult to measure bioavailable N.

33. The Sulfur Cycle

Cycling of sulfur in the environment takes place between the atmosphere (gaseous S), the hydrosphere (dissolved S) and the lithosphere (solid S), with an important component passing through the biosphere. The lithosphere (rocks and sediments) is the largest reservoir of S, followed by marine water, fresh water, the biosphere and the atmosphere (Table 33.1).

Table 33.1 Reservoirs of sulfur in the environment.

Environmental reservoir	Mass of sulfur (Gt)	Dominant form
Atmosphere	1.8×10^{-3}	SO_2
Biosphere	10*	
Fresh water	1.2×10^3	SO_4^{2-}
Marine water	1.2×10^6	SO_4^{2-}
Marine sediment	2.5×10^5	S^{2-}
Continental sediment		
Oxidised	6×10^6	SO_4^{2-}
Reduced	4.5×10^6	S^{2-}
Rocks	1.75×10^7	S^{2-}

*Estimated from a N:S ratio of 7:1 in plant tissue and a C:S ratio of 100:1 in organic matter.

Transformations in the chemical form (speciation) of sulfur bring about transfers of S between these different reservoirs. Figure 33.1 shows typical annual fluxes of S between the atmosphere, the oceans and the land. Note that these figures are in balance, with overall gains and losses in each component being equal:

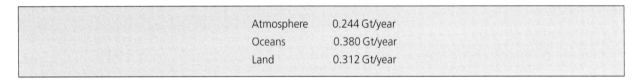

Atmosphere	0.244 Gt/year
Oceans	0.380 Gt/year
Land	0.312 Gt/year

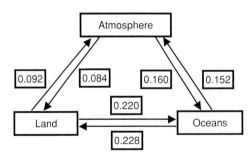

Figure 33.1 Fluxes of sulfur in the environment (units Gt/year).

Processes in the Sulfur Cycle

The main transformations in the sulfur cycle (Figure 33.2) are between reduced forms (sulfide) and oxidised forms (principally sulfate in solution and sulfur dioxide gas). In the non-biomass part of the cycle, this interchange is controlled by the redox potential (see Topic 12) and driven by microbial action. Oxidation of sulfide is brought about mainly by chemoautotrophic bacteria of the genus *Thiobacillus*:

$$H_2S + 2O_2 \longrightarrow SO_4^{2-} + 2H^+$$

Topic 33 The sulfur cycle

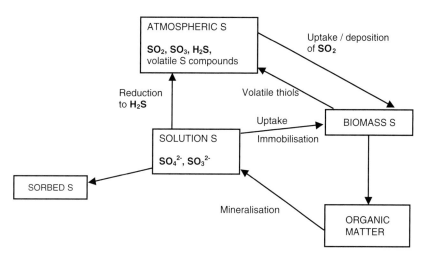

Figure 33.2 Processes in the sulfur cycle.

Reduction of sulfate is brought about by the anaerobic bacteria of the genus *Desulfovibrio* in reduced, organic environments (represented by CH_2O):

$$SO_4^{2-} + 2H^+ + 2CH_2O \longrightarrow H_2S + 2H_2O + 2CO_2$$

Sulfide (H_2S gas and metal sulfides) is the stable form only under highly reduced conditions: at E_h values greater than about -0.1 V oxidation to sulfate occurs. Thus sulfide is found only in anoxic water, soils and sediments. Hydrogen sulfide gas can escape to the atmosphere, but often the sulfide anion reacts with metal cations to form an insoluble sulfide. Iron sulfide is particularly significant, black FeS in the short term and FeS_2 (pyrite) in the long term (see Topic 18), and many of the important metal ores that are mined are sulfides (e.g. CuS, ZnS, PbS – see Topic 46).

Much of the S taken up by the biosphere organisms, mainly in the form of sulfate, is reduced within the cells for use in the building of compounds such as certain amino acids (and hence proteins, where the S—S bond is important for structural conformation) and vitamins. Some S is retained in the oxidised form, for example, sugar sulfates. These two classes of S compounds – termed carbon-bonded sulfur and sulfate esters, respectively – are recognised in humic material. Sulfur in the second category can be released by the action of sulfatase enzymes, while that in the former requires the C—S bond to be broken. A C:S ratio of less than 100 results in net mineralisation of S. Oxidation of the biomass or humified material – for example, combustion of plant tissue, burning of fossil fuels – results in oxidation of S to SO_2 and its release into the atmosphere. This is one of the causes of acid rain (see Topic 52). Some biomass organisms release volatile forms of S into the atmosphere. In marine systems the main compound released is dimethylsulfide $(CH_3)_2S$, produced by the phytoplankton. This, and the H_2S released mainly from terrestrial environments, is rapidly oxidised in the atmosphere to SO_2.

Although the lithosphere is the part of the environment containing the greatest amount of sulfur, its role is considerably less important than in the phosphorus cycle (see Topic 34). Volcanic action is a major process whereby S is transferred from the lithosphere to the atmosphere, as SO_2. In the short term, the redox changes in soils and sediments described above are significant in those environments where they operate, or when highly reduced systems are exposed to oxidising conditions; for example, coal mine waste (see Topic 46), reclaimed polders (see Topic 18). Sulfate accumulates only in arid or semi arid terrestrial environments, usually as gypsum ($CaSO_4$), which is reasonably soluble so its persistence depends on there being insufficient water present to dissolve the salt and leach away the ions. Sorption processes are relatively unimportant for sulfate, which contrasts strongly with their importance for phosphate.

Atmospheric processes involving sulfur are complex and are discussed in Topic 52.

Bioavailability of Sulfur

The immediate source of sulfur to living organisms is the sulfate anion in solution, which, as sorption of sulfate is relatively unimportant, can be considered the main bioavailable pool. The concentration of sulfate in solution is maintained primarily by the release of S from organic matter, either by sulfatase enzyme action or by breaking of the C—S bond.

34. The Phosphorus Cycle

Phosphorus is a relatively minor element in terms of its concentration in the Earth's crust, with an average concentration of about 1000 mg/kg; in soil a typical concentration is 400–600 mg/kg. It is, however, required by living organisms in relatively large amounts; for example, plants typically contain up to 0.5% P in their dry tissue, while some micro-organisms may contain up to 2.5% on a dry weight basis. Phosphorus is routinely added as fertiliser to most agricultural soils (see Topic 38).

The behaviour of phosphorus in the soil–water system (Figure 34.1) is controlled by both chemical and biologically mediated reactions:

- Weathering of mineral phosphates
- Chemisorption of phosphate
- Immobilisation and mineralisation by biomass and organic matter.

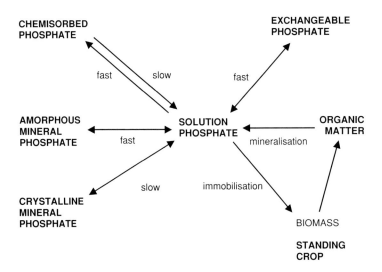

Figure 34.1 The phosphorus cycle.

Weathering of Phosphorus Containing Minerals

Solubility of phosphorus containing minerals is an important control on the behaviour and mobility of P in the environment. Crystalline mineral phosphates are highly stable and form long term reserves of P. Amorphous mineral phosphates are more reactive as they exist as coatings on other minerals (e.g. clays, oxides, etc.) and so have a high surface area. Over time, the amorphous mineral phosphates become more stable, less reactive and less bioavailable as they slowly transform to crystalline mineral phosphates. Calcium phosphates are important in slightly acidic, neutral and alkaline environments, while iron and aluminium phosphates are more important at acid pH (Table 34.1).

Table 34.1 Forms of mineral phosphate.

Crystalline mineral phosphates	
Octacalcium phosphate	$Ca_8H_2(PO_4)_6 \cdot 5H_2O$
Hydroxyapatite	$Ca_5(PO_4)_3(OH)$
Strengite	$FePO_4 \cdot 2H_2O$
Variscite	$AlPO_4 \cdot 2H_2O$
Amorphous mineral phosphates	
Dicalcium phosphate	$CaHPO_4$
Iron phosphate	$FePO_4 \cdot nH_2O$
Aluminium phosphate	$AlPO_4 \cdot nH_2O$

Phosphorus in Solution

Phosphorus in solution exists entirely as the orthophosphate anion, the exact form of which varies with pH.

$$H_3PO_4 \rightleftharpoons H_2PO_4^- + H^+ \quad pK\,2.15$$
$$H_2PO_4^- \rightleftharpoons HPO_4^{2-} + H^+ \quad pK\,7.20$$
$$HPO_4^{2-} \rightleftharpoons PO_4^{3-} + H^+ \quad pK\,12.35$$

The pK value is the pH at which 50% of the P is in each form in the three equations. $H_2PO_4^-$ is the solution species of P that dominates over much of the natural environment, up to about pH 7, while HPO_4^{2-} is dominant in alkaline systems.

Concentrations of phosphate in natural waters are extremely low because of the insolubility of P minerals, the strong affinity of phosphate for chemisorption onto oxides, and the high amount of P in the biomass and humified organic matter. For example, micromolar concentrations are typical in soil solution, which are among the highest in the hydrosphere due to the effect of plant roots and fertiliser addition. Because of this low solution concentration, little phosphate is leached from soils and sediments. Transfer of P from the land to water is primarily due to erosion of phosphate chemisorbed onto soil particles or in organic matter. Because of the low solution concentration of P in lakes and rivers, which is often the limiting factor to growth, such transfer can result in eutrophication (see Topic 42).

Exchangeable and Chemisorbed Phosphorus

Although it is possible for the phosphate anion to be held on positively charged surfaces, for example hydrous oxides at pH below their pznc (see Topic 4), this is a relatively unimportant process in the cycling of P in the environment.

Chemisorption of phosphate ions onto surfaces of iron and aluminum hydrous oxides (see Topics 4 and 8) and short-range order aluminosilicates (see Topic 1) is the main adsorption process controlling the behaviour and mobility of P in the environment (Figure 34.2). In alkaline systems, phosphate can be taken up on the surfaces of calcite ($CaCO_3$), but the distinction between sorption and precipitation is unclear.

The affinity for the surface is so high that phosphate ions are chemisorbed even onto negatively charged surfaces, despite the charge repulsion. These ions are strongly held, but can slowly return to the solution phase in response to a fall in phosphate concentration.

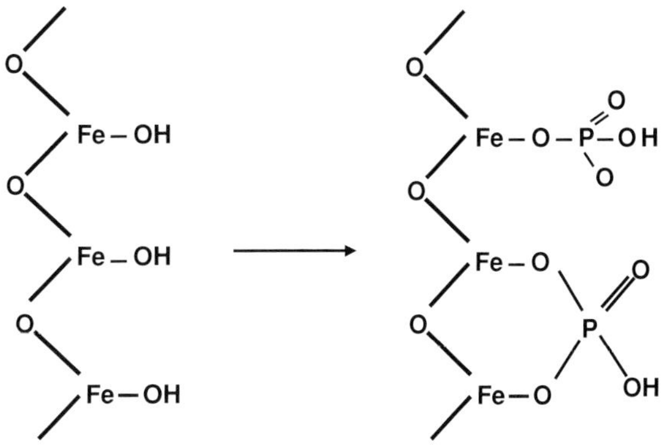

Figure 34.2 Chemisorption of phosphate onto an iron oxide surface.

Organic Phosphorus

Phosphorus containing compounds that enter environmental systems from living organisms are mainly organic phosphate esters, such as nucleic acids, phospholidids, sugar phosphates and phytates. The latter, also known as inositol phosphates, are non-aromatic ring systems with between one and six phosphate groups attached by ester linkages (Figure 34.3).

Most of these compounds are rapidly broken down in the environment by microbial action, but inositol phosphates tend to be stable in soil, possibly due to the formation of salts with Ca, Al and Fe. Penta-, tetra-, tri-, di- and mono-inositol phosphates are found, representing successive stages of microbial breakdown. Phosphorus gets incorporated into the humic material, and about 50% of the P in soil organic matter is unidentified; the rest is largely inositol phosphates and the compounds listed above, which have been released on the death of cells.

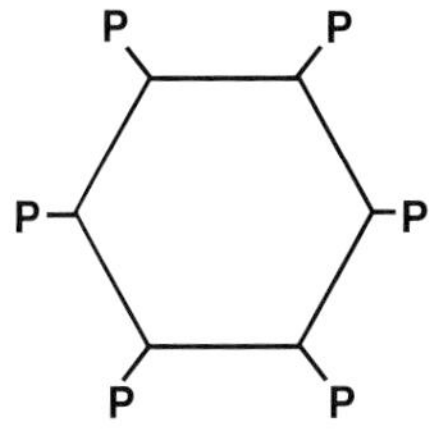

Figure 34.3 Inositol hexaphosphate (phytic acid) (P = orthophosphate).

The balance between mineralisation and immobilisation depends on the C:P ratio in the material being decomposed: the theoretical critical C:P ratio is 100:1; lower than this favours mineralisation, greater than this favours immobilisation. In reality, the ratio is often found to be closer to 200:1 because phosphatase enzymes released into the environment from microbial cells can release phosphate by breaking the ester linkage without the need to completely break down the organic molecule.

Bioavailability of Phosphorus

The immediate source of P to living organisms is the orthophosphate ion in solution. It is difficult to define unambiguously the bioavailable fraction of phosphorus, but P held in amorphous minerals, chemisorbed onto oxide and other surfaces and in organic matter, can be released in response to a decrease in solution concentration (e.g. due to plant uptake). These forms of P may be made more soluble by the action of root and microbial exudates and the acidifying effect of root and microbial respiration.

35. The Potassium Cycle

Potassium is a major element in the environment, with an average concentration in the Earth's crust of 2.6% and in soil of 0.8% (see Topic 1). It is second only to nitrogen as the element required in the greatest amounts by plants, which typically contain 1–5% K in their dry tissue. Crops annually remove from soil between a few tens of kilograms of potassium per hectare (kg K/ha) for cereals to about 100–250 kg K/ha for high demanding species such as certain grasses, potatoes, tomatoes and sugarcane. Potassium is routinely added as fertiliser to most agricultural soils (see Topic 38).

The behaviour of potassium in the soil–water system (Figure 35.1) is controlled mainly by chemical reactions:

- Weathering of K containing minerals
- Cation exchange reactions
- Fixation and release of K in certain soils.

Although the release of K during the breakdown of the organic matter is of little importance, it should be remembered that a significant fraction of the K in the system may be held in the biomass or growing crop.

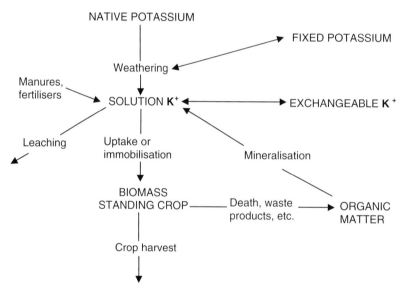

Figure 35.1 The potassium cycle.

Weathering of Potassium Containing Minerals

The main K containing minerals are feldspars (microcline, $KAlSi_3O_8$), micas (muscovite, $KAl_2(Si_3Al)O_{10}(OH)_2$, and biotite, $K(Mg,Fe)_2(Si_3Al)O_{10}(OH)_2$), and the clay mineral illite (see Topic 1). They are broken down by weathering, especially by acid hydrolysis (see Topic 2), but the rate of release of K is extremely slow. Potassium is released into the aqueous phase (soil solution, sediment interstitial water, river or lake water) in the form of K^+ ions.

Potassium in Solution

The solution chemistry of potassium is entirely as the K^+ cation. As a member of the alkali metal group (group 1 of the periodic table), K forms soluble salts and does not form stable complexes with organic matter. K^+ ions are readily taken up by plants and micro-organisms, and can be leached into surface- and ground-waters. The main control on the concentration of K^+ ions in solution, and their mobility in the surface environment, is their uptake onto ion exchange sites (see Topic 7) or fixation in some clay minerals. Both of these processes proceed at a much faster rate than the release of K by weathering (Figure 35.2).

Typical concentrations of potassium in different parts of the hydrocycle are shown in Table 35.1. Although most of the land based systems are reasonably constant in composition, reflecting variability in the geology through which the water has

Topic 35 The potassium cycle

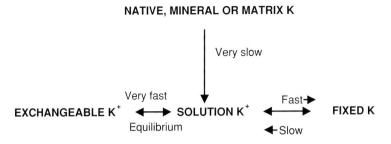

Figure 35.2 Reactions involving K$^+$ ions.

Table 35.1 Typical concentrations of potassium in the hydrocycle.

	K concentration (mg/dm^3)
Rainwater	<1
River water	1–2
Soil water	4–40
Lake water	2–800
Groundwater	4–8
Seawater	400

passed, lake water in particular is variable as there is the added effect of significant evaporation in some cases. Values at the top end of the range given in Table 35.1 are found in enclosed lake systems which act as sinks for dissolved salts.

Exchangeable and Fixed Potassium

The most important reaction governing the mobility of potassium in the environment is its uptake as K$^+$ ions onto cation exchange sites (see Topic 7). The main exchange sites are on the aluminosilicate clay minerals (Figure 35.3), with minor amounts held by sites on the humic material and hydrous oxides. Ions held on external, planar sites on clays can easily be exchanged for other ions and returned back into solution. When the K$^+$ ions are held on the surfaces between the unit layers of non-expanding clays, release of the ions back into solution is slow and they are said to be 'fixed'. Their release is governed by the ability of other cations to diffuse into the interlayer space to exchange with the K$^+$ ions, which is limited by the size of the replacing ion. As weathering of the clay mineral proceeds, the unit layers can move apart at the edge to form a so called 'wedge site' from which exchange of potassium is easier.

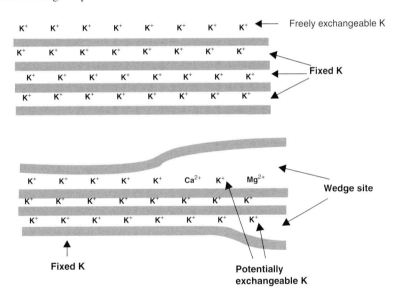

Figure 35.3 Potassium ions held at clay surfaces in freely exchangeable, potentially exchangeable and fixed forms, (━━━ represents a 2:1 unit layer of a non-expanding clay mineral).

Bioavailability of Potassium

The form of K taken up by plants and micro-organisms is the K$^+$ ion in solution. Freely exchangeable K can be readily brought into solution by ion exchange, and so is considered bioavailable; fixed K is less bioavailable and mineral K is unavailable. Bioavailability of K in soil is measured by extracting the soil with a dilute solution of another salt (e.g. ammonium nitrate). The cation of the extractant (e.g. NH$_4^+$) exchanges with K$^+$ held on cation exchange sites, bringing it into solution. The concentration of K in the extract can then be measured and related back to the concentration of exchangeable (bioavailable) K in soil.

Pollution

SECTION I

36. Sources of Pollution

Pollution occurs due to an anthropogenic (human) activity that causes either:

- A concentration of an element or compound to a level that can disrupt, or be toxic to, biological activity, or
- Production of a persistent synthetic compound that can disrupt, or be toxic to, biological activity.

The main effects of many sources of pollution are on human health, and this is often the driving force behind mitigation and clean up procedures. But the health of the environment is also important and can be affected by pollution so that the functioning of environmental processes is impaired. A good example of these different focuses is the release of sulfur dioxide due to the burning of fossil fuels. The initial focus was on human health, especially respiratory diseases such as asthma where a link was made to the emissions of smoke and SO_2 by coal burning. In the UK, the London smogs of the early 1950s led to the Clean Air Acts of the 1950s and 1960s, which did much to improve air quality and hence impacted on human health. In the 1970s and 1980s, however, the focus switched to the emissions of SO_2 by power stations and the formation of acid rain, which affected soils and waters. The reduction of these emissions was driven by concerns for the environment.

The main activities that lead to pollution are shown in Table 36.1.

Table 36.1 Major sources of pollution.

Industry	Release of heavy metals – toxicities
	Release of organics – toxicities
	Emission of smoke and gases – health effects
	Heated water – promotion of biological activity
Power production	SO_2 emissions – acidification, health effects
	CO_2 emissions – greenhouse gas
	Nuclear waste – health effects
	Nuclear accidents – health effects
Mining and smelting	Release of heavy metals – toxicities
	Acid mine drainage – acidification
	SO_2 emissions – acidification, health effects
Transport	NO_x emissions – acidification
	CO_2 emissions – greenhouse gas
	Pb emissions – health effects
	Platinum group elements – effects unknown
	Particles – health effects
Domestic	Organic wastes to landfill – leachate and methane production
	Sewage disposal – heavy metals and organics
Agriculture	Nutrient leaching and enrichment – eutrophication of surface waters
	Nitrous oxide emissions – greenhouse gas
	Ammonia volatilisation – acidification
	Slurry and silage effluents – contamination of rivers
	Methane production – greenhouse gas

The source of pollution can be either a *point source* or a *diffuse source*. A point source of pollution comes from a single, identifiable origin and its effects are usually seen in a relatively restricted area. Figures 36.1 and 36.2 show an example of a point source of pollution entering a river and its effects on two invertebrate species. The pollutant has entered the river somewhere between sample points 1 and 2. The increase in suspended solids shows that the pollutant is in particulate form and not in solution. The decrease in dissolved oxygen suggests that it is organic because it is a carbon source for heterotrophic organisms, whose numbers increase to exploit this resource, and so the demand for oxygen by these aerobic organisms strips the water of its dissolved oxygen. This could be, for example, an overflow from a sewage works or landfill site, where in both cases organic wastes are being broken down. Downstream from sample point 2 the suspended solids content falls as micro-organisms break down the pollutant. The dissolved oxygen content rises because the numbers of micro-oganisms will decrease as the carbon source is used up, and hence the oxygen demand falls. This is called a *self-purification process* and the changes in dissolved O_2 an *oxygen sag curve*.

These changes have consequences on the living organisms in the river. Stoneflies are particularly sensitive to oxygen content of the water and their numbers fall to zero in the most polluted stretch of the river. Bloodworms, on the other hand, have the ability to scavenge oxygen from water, even when the dissolved content is low; they therefore have an ecological advantage

Topic 36 Sources of pollution

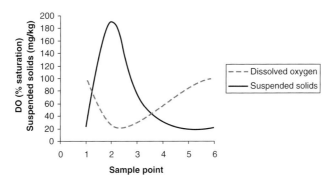

Figure 36.1 Effect of a point source of pollution on a river.

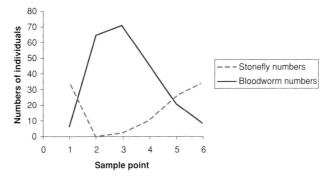

Figure 36.2 Effect of the pollution on two invertebrate species.

in this situation and so their numbers increase. When the dissolved oxygen content of the water rises again, stonefly numbers increase and bloodworms decrease.

A diffuse source of pollution does not come from one specific source and has an effect over a large area. For example, nutrient enrichment of lakes and rivers results from leaching of nitrate in solution and erosion of soil particles with phosphate bound to them into surface waters where the nutrients can cause algal blooms. Water flowing from areas of coniferous forestry tends to be of low pH and so has an acidifying effect on lakes and rivers. These inputs come from a wide area, which is defined by the river catchment (Figure 36.3). Acidification of lakes and rivers is a result of the emission of gases that have been transported over long distances, so that their effects are seen far away from the original sources.

Splitting sources of pollution in this way, although commonly done, may not always be appropriate. A single motor car acts as a point source, but its emissions of gases and particulates influence a wide area. An individual refrigerator can be a point source of chlorofluorocarbons (CFCs), but their effect on depletion of stratospheric ozone is global in its action. In such cases a definition of a point source could be based on the ability to take control measures to prevent pollution. Thus action can be taken to prevent a single car or refrigerator acting as a point source of pollution.

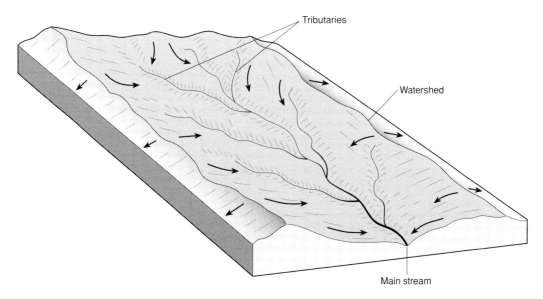

Figure 36.3 Pathways of diffuse pollution in a river catchment.

37. Pesticides

From the beginnings of agriculture, farmers have sought to control pests and diseases in their crops. The Sumerians used sulfur to control insect pests 5000 years ago, while 2500 years ago the Chinese were using mercury and arsenic compounds. Crop rotation to reduce the carry-over of diseases has also been used for thousands of years. Modern industrialised agriculture uses a large array of chemicals to control pests and diseases in crops and stored food. We also use a wide range of pesticides in our homes and gardens.

In broad terms a pesticide is a chemical that kills pests, but more specifically:

- Insecticides kill insects
- Fungicides kill fungi
- Herbicides kill weeds
- Molluscicides kill molluscs such as slugs
- Rodenticides kill rats and mice.

Soil sterilants, wood preservatives, plant growth regulators, bird and animal repellents, and household products such as bleaches and surface sterilants are also included. The world market for pesticides was worth approximately £6 billion in 2000.

Farmers use pesticides to protect crops from insect pests, weeds and fungal diseases while they are growing, and to prevent fungi, rats, mice, flies and other insects from contaminating foods whilst they are being stored. Despite the current use of pesticides, it is estimated that worldwide crop losses of 35% occur in the field due to insects, fungi and weeds, while a further 20–30% is lost post-harvest due to spoilage by rodents, insects and fungi. Between 1845 and 1849 fungal disease devastated the potato crop in Ireland resulting in one million deaths.

Pesticides are an important weapon against diseases, particularly insect vectored diseases such as malaria, yellow fever, sleeping sickness, river blindness and typhus. The Black Death, caused by a bacterium spread by fleas, resulted in the death of 25% of the population of Europe in the mid 14th century. As recently as the 20 year period up to the end of WWI in 1918, it is estimated to have caused 10 million deaths worldwide. Today 40% of the world's population are exposed to malaria, and estimates of the number of deaths resulting from malaria are 1–2 million per year.

While synthetic chemicals have produced enormous benefits in food production and quality, and in the control of diseases, they also pose a number of serious problems. It is estimated that 90% of the pesticides used miss the target species. As a result, pesticide residues spread in the environment, beneficial organisms may be killed and there are concerns about the levels of pesticide residues in food. There is also the problem of the development of resistant strains of pest and diseases. The World Heath Organization estimates that 25 million people suffer symptoms from pesticide poisoning each year. Predominantly these are acute effects caused by occupational exposure resulting from poor application practices and inadequate protective clothing.

Environmental Fate of Pesticides

The environmental fate and impact of a pesticide applied to a growing crop is a combination of its stability and dispersal. As little as 10% of the applied herbicide may reach the target weed seedling in Figure 37.1, either by foliar or root uptake. Depending on its chemical properties, the residual chemical may volatilise into the atmosphere or leach into the aquatic environment or groundwater. Critical to the dispersal of a chemical is its binding by soil. Retention within the soil profile allows decomposition *in situ* and prevents dispersal. Breakdown may occur in the plant or following uptake by animal species, as well as by micro-organisms in soil and water. Some pesticides are susceptible to photo-decomposition.

Some pesticides have the property of *bioaccumulation* – building up in the food chain. This is particularly true of chemicals with high fat solubility such as DDT, which shows a concentration factor of 10^7 between the water and fish-eating birds at the top of the food chain in Table 37.1.

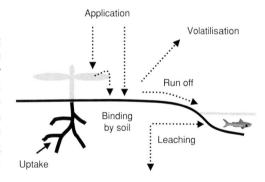

Figure 37.1 Dispersal of a pesticide in the environment.

Dispersal and stability are therefore closely linked (Figure 37.2). Very stable chemicals become widely dispersed, even if they are not very mobile in the environment, because of the time available for dispersal. The impact may depend on the properties of the degradation products as well as the parent chemical. The ideal properties of a pesticide include target specificity in its effect, low water solubility, low volatility and strong binding by the soil to minimise dispersal, and sufficient stability to be effective, after which it should degrade rapidly to harmless products. The balance between these properties depends on the

Topic 37 Pesticides

Table 37.1 Bioaccumulation of DDT in an aquatic food chain.

	DDT concentration (mg/kg)
Fish eating birds	25
Large fish	2
Small fish	0.5
Plankton	0.04
Water	0.000003

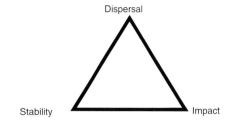

Figure 37.2 Environmental dispersal and impact of pesticides.

application. For example, with pesticides applied to food products, low persistence and low residues are a priority, while for an insecticide used in mosquito control, greater persistence and therefore less frequent application is an advantage.

Environmental Testing of Pesticides

Developing a new pesticide requires extensive toxicity and environmental testing for the following:

- *Acute toxicity* towards non-target species and effects such as mutagenicity, carcinogenicity, teratogenicity and hormonal and endocrine disrupting effects
- *Metabolism studies* to establish the breakdown pathways in plants, animals soil and water
- *Physiochemical testing* to determine properties such as water and fat solubility, volatility, behaviour in soil and water
- *Residue studies* to determine the levels of the chemical and its breakdown products in treated food crops in different agricultural systems
- *Ecological testing* to determine the impacts of non-target populations of soil micro-organisms, plants, insects, earthworms, birds, mammals and fish in test plots and the wider environment.

Only when all of these hurdles have been successfully cleared will a new chemical be licensed for sale.

DDT and the Organochlorines

Dichlorodiphenyltrichloroethane (DDT) (Figure 37.3) and the related organochlorine class of insecticides provide a classic example of the promise and problems associated with the use of pesticides. DDT was first synthesised in 1873, but its insecticidal properties were not discovered until 1939. It was widely used during and just after WWII to control the spread of typhus in refugee populations by controlling the human body louse. The World Health Organization, which, in 1955, started a programme to control the *Anopheles* mosquito that spreads malaria estimated that the programme saved 15 million lives in its first 10 years. DDT was seen as the ideal chemical for the control of insect vectored diseases because of its low cost, low mammalian toxicity and stability. When in the 1950s it was released for use as an agricultural insecticide it was seen as the start of a new era of crop production with greatly increased and better quality yields.

Figure 37.3 Structure of DDT.

Widespread use, however, soon resulted in the appearance of problems. Insect pest populations, including malaria carrying mosquitoes, started to develop resistance requiring larger and more frequent applications, and eventually control failed. This was addressed by development of related compounds such as benzene hexachloride, aldrin and dieldrin. The wide spectrum impact of DDT meant that many non-target insect populations were affected and in turn the organisms higher in the food chain such as birds which fed on them.

The chemical stability and fat solubility of DDT caused further problems. The biological half-life of DDT is about 8 years and with continued ingestion DDT builds up within the animal over time. Bioaccumulation particularly affected the predators at the top of the food chain. Numbers of birds of prey decreased greatly as the accumulated DDT in their bodies interfered with calcium metabolism, resulting in the formation of very thin eggshells and low breeding success. The persistence of DDT resulted in it becoming widespread in the environment together with its major decomposition product DDE (dichlorodiphenyltrichloroethylene), which is more persistent than DDT, though it is subject to photodecomposition, and is an endocrine disrupting ('gender bender') agent.

Rachel Carson's book *Silent Spring* published in 1962 proved to be a turning point. The title referred to the absence of songbirds but highlighted the growing concerns about the indiscriminate use of persistent pesticides. Banned in the USA in 1971 and in the UK in 1974, DDT is now banned virtually worldwide. However, it is estimated that there are still several million tonnes of DDT dispersed in the environment, mostly dissolved in the oceans where concentrations as low as 10 parts per billion can significantly reduce photosynthesis by the marine plankton. Dieldrin and aldrin were finally banned in UK in 1989, but benzene hexachloride is still available in gardening shops. DDT provides an example of the unresolved dilemma of how pesticides can play a vital role in public health and in supplying food to an increasing world population while minimising the costs to the environment.

38. Fertilisers in Agriculture

Fertiliser Use in Agriculture

The main plant nutrient elements added to agricultural soils as fertilisers are nitrogen, phosphorus and potassium (N, P, K). Figure 38.1 shows the changes in total fertiliser use in the UK in the period 1850–2000.

The main feature is the rapid increase in the use of nitrogen during the second half of the 20th century, specifically to increase agricultural production. The amount of N used has declined over recent years, due partly to an increasing awareness of the problems of eutrophication (see Topic 42) and contamination of some drinking water supplies (see Topic 40) as a result of nitrate leaching, but also to the removal of land from agricultural production (set-aside). Other macronutrients (S, Ca, Mg and Cl), and micronutrients (B, Cu, Fe, Mn, Mo and Zn) tend to be added as fertiliser only if there is a proven deficiency requiring correction. However, in some places S deficiency is being increasingly found as a result of improvements in air quality (see Topic 52).

Crops respond to fertilisers as shown in Figure 38.2. Most soils have some available nutrient that crops can use, so there will be a yield even with no fertiliser added (Y_0). As fertiliser additions increase, there will be a response in crop yield up to a maximum value, which is the optimum fertiliser application. Beyond that a decrease in yield will occur due either to direct toxicity or to the increase in soluble salts in the soil. Response curves will differ depending on crop requirement for a particular nutrient and on soil properties, but the overall shape remains the same.

Fertiliser additions are determined for N on the basis of the previous crop grown, its demand for N and therefore the likely residual N carried over into the next growing season. For P and K, measurements are made using extractants that are thought to remove the bioavailable pool from soil (e.g. in England and Wales, 0.5 M $NaHCO_3$ and 1 M NH_4NO_3 are used for P and K, respectively) and a knowledge of crop responses on particular soils based on field experience.

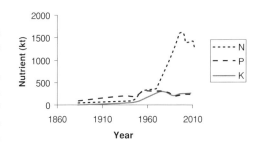

Figure 38.1 Total fertiliser use in the UK. *Source*: Fertiliser Manufacturers Association and Defra.

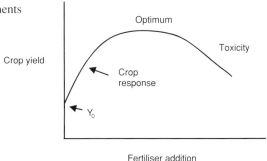

Figure 38.2 Typical crop yield response to fertiliser addition.

Common Fertiliser Types

Fertilisers can be applied as single element (or straight) fertilisers, or in combination, e.g. NPK compound fertilisers.

Nitrogen Fertilisers

Natural nitrogen containing minerals are rare, so almost all (∼97%) nitrogen fertilisers are produced from ammonia made industrially by the Haber–Bosch process. This reacts nitrogen from the atmosphere with hydrogen obtained from a feedstock such as natural gas

$$N_2 + 3H_2 \longrightarrow 2NH_3$$

This is carried out at high temperature and pressure using a catalyst, and so is a high energy demanding process (current estimates are about 25–30 MJ per kilogram of NH_3 produced).

Table 38.1 shows the main straight nitrogen containing fertilisers used in western Europe. These account for almost 80% of nitrogen fertilisers used in this area, with the other 20% being compound fertilisers with P and/or K.

Ammonium nitrate and calcium ammonium nitrate (which contains at least 20% calcium or magnesium carbonate) are the dominant forms of N fertiliser in Europe. In the UK, ammonium nitrate accounts for almost 70% of straight N fertilisers used. The reason for this popularity is that half the nitrogen is in the form of nitrate, which is immediately available to crops, but readily lost by leaching (see Topics 39 and 40), and half as ammonium, which can be held in the soil by ion exchange (see Topic 7) and converted to nitrate over time by the nitrifying bacteria (see Topic 32). In the rest of the world, urea is much more commonly used, especially in developing countries where it accounts for almost 70% of N fertiliser use. This is partly economic (higher N content means more N per unit weight) and partly due to ready availability and ease of application. However, the potential for loss of N due to ammonia volatilisation is considerable.

Topic 38 Fertilisers in agriculture

Table 38.1 Use of nitrogen fertilisers in western Europe.

	Nitrogen content (%)	Approx. amount used 2001/2 (kt)
Calcium ammonium nitrate	25–28	2600
Ammonium nitrate	33.5–34.5	2300
Urea	46	1800
Liquid urea/ammonium nitrate	28–32	1300
Ammonium sulfate	21	250
Anhydrous ammonia	82	80
Others (mainly nitrates)	Variable	400

Source: European Fertiliser Manufacturers Association.

Phosphorus Fertilisers

The source of all phosphorus fertilisers is rock phosphate, which is a sedimentary rock containing apatite ($Ca_5(PO_4)_3(OH)$) or fluorapatite ($Ca_5(PO_4)_3(F)$). The apatites in *ground rock phosphate* (25–40% as P_2O_5 – the way in which the P content of fertilisers is traditionally presented) are poorly soluble, and so used only as long term supplies of P in low production agriculture such as upland grazing. To increase the solubility, rock phosphate is treated with either sulfuric acid to produce *superphosphate* (~20% P_2O_5) or phosphoric acid to produce *triple superphosphate* (47% P_2O_5). Approximately 66% of current global production of rock phosphate comes from the USA, Tunisia and China.

Potassium Fertilisers

Potassium fertilisers also come from the mining of a natural deposit; in this case salt deposits laid down millions of years ago in hot climates when seawater lagoons were isolated and the water evaporated off. The main sources are in Canada, Russia, Belarus and Germany. The most common salt used is potassium chloride (also called muriate of potash), which is about 95% of K fertilisers used, and this contains 60% potassium expressed as K_2O. The other K salt used for fertilisers K_2SO_4, which is more expensive but does add in a second macronutrient, sulfur.

Compound Fertilisers

Compound fertilisers containing a combination of N, P and K are made either by mixing granules of the fertilisers described above, or by forming new granules from solutions of the N, P and K salts. This is the most common form in which P and K fertilisers are used. Compound fertilisers are labelled with the percentage of each component as N, P_2O_5 or K_2O, so an 18:12:10 compound fertiliser contains 18% N, 12% P_2O_5 and 10% K_2O. Under EU regulations the percentage of N as nitrate and ammonium, percentage water soluble P_2O_5 and K_2O (and equivalent values as P and K) must be included in the labelling on bags of fertiliser.

Organic Fertilisers

Organic wastes are used as a source of nutrients, especially nitrogen (Table 38.2). The main wastes used are animal manures (relatively solid, mix of dung, urine and bedding such as straw) and slurries (relatively liquid, mix of dung and urine) from agriculture, and sewage sludge from domestic and industrial wastes (see Topic 41). Most additions to soil are as animal wastes as these are produced on-farm. There are restrictions to the amount of sewage sludge that can be added to soils because of the heavy metal content, which could result in toxicity. Nitrogen in these wastes is predominantly in the NH_4^+ form due to the urea and uric acid content of the material.

Table 38.2 Typical nutrient content of organic wastes.

	Total N content (% in dry matter)	Total P content (% in dry matter)	Total K content (% in dry matter)
Cattle manure	1–3	0.5	2
Cattle slurry	3–5	0.8	4
Pig manure	2–3	1	1.3
Pig slurry	5–10	1.6	2.4
Poultry manure	3–5	1.3	1.8
Poultry slurry	3–7	1.8	2
Sewage sludge	3–5	1.2–1.5	

Environmental Impact of Fertiliser use

The main impacts of nitrogen fertilisers on the environment are:

(1) Nitrate leaching, leading to health risks in drinking water (see Topic 40) and eutrophication risks in surface waters (see Topic 42)
(2) Volatilisation of ammonia, which can lead to acidification when deposited in soils and waters (see Topic 52)
(3) Release of nitrous oxide (N_2O) by denitrification under moderately anaerobic conditions (see Topic 12), which is a potent greenhouse gas (see Topic 49).

39. Nitrate Leaching

The negatively charged nitrate anion is very poorly held in the predominantly negatively charged soils of temperate regions (see Topic 7) and consequently it is easily leached through the soil profile when drainage of water occurs. Although nitrate is formed in unfertilised soils through the processes of ammonification and nitrification (see Topic 32), the primary source of nitrate leaching into groundwater and rivers is from fertilised agricultural soils.

Table 39.1 shows that of the major constituents of an NPK compound fertiliser, the positively charged ions such as NH_4^+ and K^+ are held in the soil profile by cation exchange and phosphate is very strongly held by adsorption on to iron oxides. There is, however, no mechanism to hold nitrate. Ammonium is quickly oxidised to nitrate by the nitrifying bacteria, so all the N content of fertilisers is vulnerable to leaching.

Table 39.1 Holding of constituent ions of a compound fertiliser by soil surfaces.

Ion	Charge	Surface binding mechanism
Ammonium	+	Cation exchange
Nitrate	−	Not held
Phosphate	−	Adsorption by oxides
Potassium	+	Cation exchange

Water Balance and Nitrogen Economy

For nitrate leaching to occur, rainfall must exceed evaporation, the soil be at field capacity so that there is downward movement of water, and nitrate must be present in the soil profile (Figure 39.1).

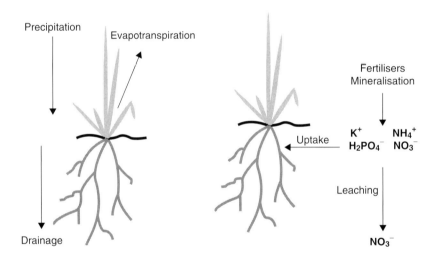

Figure 39.1 Leaching of nitrate.

Fertiliser N is usually applied in the spring and further inorganic N may be released by mineralisation of soil organic matter once the soil profile warms sufficiently. Nitrification of ammonium results in nitrate becoming the major form of inorganic N in the profile. Early fertiliser applications while the soil is close to field capacity are vulnerable to leaching in the event of subsequent heavy rainfall. In the summer months there are growing plants with well developed root systems taking up nitrate, so there is a high demand for nitrate. The water balance shows an excess of evapotranspiration over rainfall, so that the soil profile gradually dries out and a soil moisture deficit develops. Rainfall events partly rewet the soil profile, but there is no drainage of water and no leaching of nitrate occurs (Figure 39.2).

By the autumn, arable crops will have been harvested so there are often no plants to take up nitrate from the soil profile. Residual nitrate from excess fertiliser will remain in the profile and may be added to by mineralisation of soil organic matter and nitrification while soil temperatures remain sufficiently high. Even at temperatures just above freezing, these processes will continue at a slow rate. Consequently, nitrate accumulates in the soil profile. In the autumn and winter, evapotranspiration is greatly reduced, so rainfall exceeds evapotranspiration, the soil rewets to field capacity and drainage commences. The initial drainage produces a flush of nitrate leaching in the late autumn–early winter period. In a wet winter all the nitrate may be leached from the soil profile, so nitrate concentrations in the leachate fall in the late winter–early spring period (Figure 39.3).

Topic 39 Nitrate leaching

Summer months

Water balance
 Evapotranspiration > rainfall
 Soil profile dries — No drainage
 Soil moisture deficit develops

Nitrogen balance — *No leaching*
 Growing plant
 Nutrient uptake — Low soil nitrate
 High nitrate demand

Figure 39.2 Water and N balances in spring and summer months.

Winter months

Water balance
 Rainfall > evapotranspiration
 Soil profile wets — Drainage occurs
 Soil returns to field capacity

Nitrogen balance — *Nitrate leaching*
 Plant harvested, no uptake
 Excess fertiliser residue in profile — Nitrate accumulates
 Warm soil, mineralisation and nitrification

Figure 39.3 Water and N balances in autumn and winter months.

Farming Systems and Crop Type

Table 39.2 Nitrate leaching risk of different crop types.

Crop type	N leaching risk
Spring sown arable crops	High
Autumn sown arable crops	High
Low nitrogen permanent pasture	Low
High N fertiliser input grass	High
Grass/clover mixtures	High
Ploughed out grass	Very high
Bare fallow	High

The influence of farming systems and crop type on the level of nitrate leaching depends partly on the levels of fertiliser N used and partly on the growth cycle of the crop (Table 39.2).

Spring sown arable crops have a well developed root system for only a short period of the year, making them vulnerable to nitrate leaching both at the beginning of the growing season and post harvest. The benefit of autumn sown crops is that the root system starts to develop in the autumn and residual nitrate in the soil profile from the previous cropping year will be taken up, so reducing nitrate leaching in the critical autumn–winter period. Early harvested arable crops are vulnerable to leaching post harvest due to residual fertiliser nitrogen and nitrate released by mineralisation of soil organic matter. For this reason, if the next crop is spring sown it is desirable to plant a temporary winter cover crop that will utilise the soil profile nitrate but which will be cultivated in as a green manure before the spring crop is planted (Figure 39.4).

An established grass crop has a well developed root system throughout the year and nitrogen uptake will occur even when soil temperatures are close to 0°C. Where low levels of nitrogen fertiliser are used, soil nitrate concentrations remain low and the potential for nitrate leaching is low. As nitrogen fertiliser inputs are increased, the risk of higher residual levels of nitrate in the soil profile in the autumn increases and consequently the nitrate leaching potential increases. A high clover content in a grass/clover sward similarly increases the nitrogen economy of the soil, leading to increased nitrate leaching. Ploughing up grass leads to a large flush of nitrogen mineralisation, because soil organic matter and crop residues rapidly decompose producing high soil profile nitrate levels and, in the absence of an established autumn–winter cover crop, very high nitrate leaching potential. Bare fallow leads to a high nitrate leaching potential in the autumn because, in the absence of a crop to take up nitrate from the soil, nitrate accumulates as a result of the decomposition of nitrogen crop residues and soil organic matter.

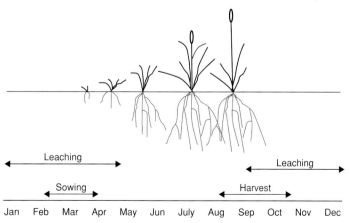

Figure 39.4 Seasonal vulnerability to nitrate leaching for a spring sown crop.

40. Nitrate in Drinking Water

Drinking water is abstracted from surface water (rivers, lakes and reservoirs) or from groundwater (chalk, limestone and sandstone aquifers). The relative importance of these two sources varies across the UK, with surface water being more important in the north and west and groundwater becoming of greater significance in the east and south-east (Figure 40.1).

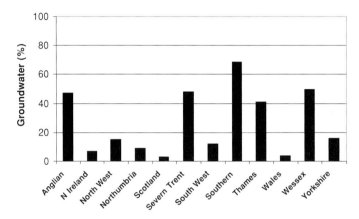

Figure 40.1 Percentage of drinking water supply coming from groundwater.

As discussed in Topic 39, nitrate leaching to both surface and groundwater is predominantly (over 70%) from agricultural fertilisers, and consequently the nitrate concentration in the drinking water supply depends on the agriculture of the region. The climate and soils of the UK mean that arable farming (e.g. cereal growing; see Figure 40.2) and high nitrogen fertiliser use is concentrated on lowland farms, predominantly in the east and south-east, while in the wetter upland areas in the west and north more extensive, low N input grazing systems predominate. Consequently, despite the lower rainfall, nitrate leaching is much greater in the east and south-east, leading to higher levels of nitrate in the drinking water supply.

The distribution of climate, types of agriculture and aquifer rocks across the UK, mean that in the low fertiliser input areas in the west and north drinking water is abstracted predominantly from surface water, while in the higher fertiliser input areas of the east and south greater reliance is placed on groundwater sources. The vulnerability of these aquifers to nitrate pollution is of concern, as contamination would have long term implications for the use of water resources.

Figure 40.2 Predominant cereal growing areas in Britain.

Toxicity of Nitrate

Nitrate (NO_3^-) has relatively low toxicity; however, nitrite (NO_2^-) is much more toxic. This is reflected in the higher median lethal dose (LD_{50}) value for sodium nitrate (1270 mg/kg) compared with sodium nitrite (180 mg/kg) and the EU limits for nitrate and nitrite in drinking water (Council Directive 98/83/EC on the quality of water intended for human consumption). The total nitrate plus nitrite expressed as [nitrate/50] + [nitrite/3] should be less than unity, with the individual limits being nitrate 50 mg/L and nitrite 0.5 mg/L.

The potential risks of high levels of nitrate are *methaemoglobinaemia* and *gastric cancer*.

Methaemoglobinaemia

Methaemoglobinaemia is a condition in which the iron atom in the haemoglobin molecule (Figure 40.3) in the red blood cells is oxidised from the Fe^{2+} state (which is the normal oxidation state and which allows the blood to transport oxygen) to the

Fe^{3+} state (which is inactive in oxygen transport). In a healthy individual about 3% of the iron is in the Fe^{3+} state, but it is reduced back to the active form enzymatically. Oxidation of a higher percentage of the iron is characterised by chocolate coloured blood, reduced blood oxygen levels and a blue tinge to the lips and skin; it is potentially fatal.

Nitrate induced infantile methaemoglobinaemia (*blue baby syndrome*) primarily affects very young infants (less than 6 months old). Their stomach is much less acid than in an adult, permitting the growth of bacteria which reduce dietary nitrate to nitrite. The nitrite then interacts with the haem group, oxidising the central Fe atom and converting haemoglobin to methaemoglobin. This is most likely to occur when the infant is bottle fed with untreated well water with bacterial contamination together with high nitrate concentration. Despite the concerns of the potential risk of methaemoglobinaemia, there have been no cases in the UK since 1972 and no cases have been linked to the mains water supply.

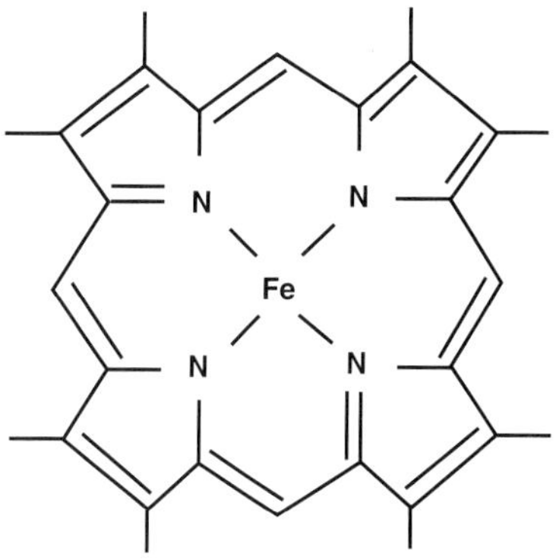

Figure 40.3 The iron atom at the centre of the haem group of the haemoglobin molecule.

Gastric Cancer

The suggested mechanism for the supposed increased risk of gastric cancer resulting from high nitrate concentration in drinking water is the formation of nitrosamines by the action of nitrite on dietary amines. The majority of nitrosamines are known to cause cancer in experimental animals, and consequently there could be a risk of cancer in man.

This mechanism also requires that the nitrate is first reduced to nitrite, which is unlikely in the more acid adult stomach or in treated mains water. Under acid conditions amines react with nitrite to form a variety of products: primary aliphatic amines decompose to nitrogen gas, while primary aromatic amines react to form a diazonium salt. Secondary aliphatic amines and secondary and tertiary aromatic amines react to form a nitrosamine:

$$\begin{array}{c} R_1 \\ \diagdown \\ R_2 \diagup \end{array} NH + HNO_2 = \begin{array}{c} R_1 \\ \diagdown \\ R_2 \diagup \end{array} N-N=O + H_2O$$

Tertiary aliphatic amines form a salt.

Although there is a theoretical risk, the epidemiological evidence is considered not to support the view that there is a real risk of increased cancer incidence linked to nitrate in drinking water.

The water companies have a legal obligation to supply drinking water which meets the 50 mg/L limit and therefore have been making the necessary investments to provide either blending systems (to mix water from different sources) or denitrification equipment in order to supply water which complies with the limit. The second area of concern, i.e. that of environmental harm, is less clear cut. The main risk from high nitrate levels is 'eutrophication' (see Topic 42). Phosphorus or nitrate may be the limiting factor in fresh water, but nitrate is the limiting factor in marine eutrophication in estuaries and coastal waters such as the North Sea.

The Nitrate Directive and Nitrate Vulnerable Zones

In 1991 Europe adopted the *Nitrates Directive* (91/676/EC), which is an environmental measure designed to reduce water pollution by nitrate from agricultural sources. As nitrate pollution from agriculture is a diffuse source of pollution (see Topic 36) the approach has been to identify areas where nitrate pollution is already, or could become, a problem for drinking water quality or environmental water quality (*nitrate vulnerable zones*) and to limit nitrogen fertiliser use within these NVZs.

The criteria used to define the NVZs were areas where surface or groundwater nitrate concentrations currently, or potentially could in the future, exceed 50 mg/L, and fresh water bodies, coastal or marine environments which are, or could become, eutrophic. Under the 2002 assessment, 55% of the land area of England plus the east coast arable areas of Scotland and a small area of Wales were designated as NVZs. From December 2002 farmers within these zones have been required to follow best practice in the use and storage of fertiliser and manure. Outside the zones this best practice is encouraged. There are four main requirements:

(1) Inorganic nitrogen fertiliser application limited to crop requirements, after allowing for residues in the soil and other sources of plant available N
(2) There is a maximum limit to total nitrogen applications in the form of organic manures
(3) On sandy or shallow soils animal manures may not be spread over the autumn–winter period
(4) The keeping of detailed records on the use of organic manures and nitrogen fertilisers.

41. Sewage Treatment

Water usage in UK is typically 150 litres per person per day (Table 41.1). While toilet flushing is a major use, baths and showers and usage by expanding numbers of kitchen appliances such as washing machines and dishwashers is increasing, despite progress in making them more efficient in water use. The majority of this water is discharged into the sewer system.

Allowing for industrial discharges to the sewer system approximately doubles the volume of sewage that must be treated. Separate collection of relatively clean water from roof and street drainage is frequently poor, resulting in further loading of the sewer system which is particularly severe in wet weather.

Table 41.1 Typical water use in the UK.

Water use	Quantity (litres per person per day)
Shower and bath	50
Flushing toilet	43
Washing clothes	37
Washing up	12
Drinking and cooking	5
Gardening, car washing	1

The Nature of Sewage

Even in dry weather the sewage arriving at the sewage treatment works of a major city is a very dilute material: dirty brown water with 0.1% total solids (about half of this is particulate and colloidal and half dissolved material). The solids fraction is predominantly organic (70%), originating from domestic sewage, but also includes such things as soap, detergents and oil, with a smaller inorganic component (30%) comprising grit, largely from street drainage and food washing, and dissolved salts. Sewage also contains pathogenic micro-organisms, and possibly toxic organic and inorganic chemicals from industrial discharges. As a result of the decomposable organic fraction, untreated sewage has a high biochemical oxygen demand (BOD) of 300–400 mg/L. It is an extremely polluting material which must be treated before it can be safely returned to the hydrological cycle.

Objectives of Sewage Treatment

Initially, in the 19th century, the objectives were to reduce the risks of disease and the smell associated with using the rivers in the major cities as open sewers. Nowadays, the minimum treatment is designed to produce an effluent that will have a minimal impact on the environment (rivers in particular) and in some cases, after dilution and natural self purification, to produce potable water. The major aims are to remove particulate and dissolved decomposable organic matter and pathogenic organisms. The 30:20 standard, reducing the suspended solids and BOD_5 (see below) to 30 mg/L and 20 mg/L, respectively, is still widely used. Depending on the environment to which the treated effluent is discharged, a secondary aim may be to remove plant nutrients such as nitrogen and phosphorus, or toxic chemicals.

Biochemical Oxygen Demand (BOD)

The BOD of an effluent is a bioassay of the consumption of oxygen by the heterotrophic micro-organisms as the organic matter decomposes. The value is expressed as milligrams of oxygen consumed per litre of effluent. The standard test measures the oxygen consumed over a 5 day incubation at 20°C in the dark (to prevent photosynthesis) giving a BOD_5 value. Water saturated with air at 1 atm and at 20°C contains 9.07 mg/L of oxygen, so the enormous pollution potential of the wastes in Table 41.2 can be seen. A very large dilution ratio (up to 3000) would be needed to prevent the receiving water becoming anaerobic.

Table 41.2 Typical BOD_5 values.

Sample	BOD_5 (mg/L)
Pig slurry	25000
Paper mill waste	25000
Vegetable processing waste	15000
Abattoir waste	2500
Domestic sewage	<500
Treated sewage effluent	<20
Unpolluted river	<2

Sewage Treatment

Sewage treatment is a complex multistage process involving a combination of sedimentation, biological digestion and chemical treatment to produce a liquid effluent suitable for discharge into a river with most of the highly polluting material concentrated in the sludge. It is a flow-though system with a total residence time of just a few hours (Figure 41.1).

Preliminary Screening and Grit Removal

Once the incoming sewage has been screened to remove large pieces of debris, its flow rate is slowed so that grit particles (sand sized material) sediment out.

Topic 41 Sewage treatment

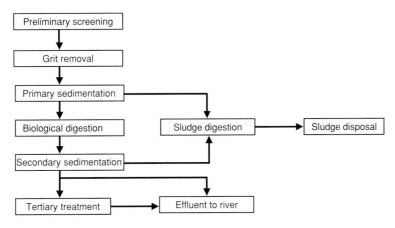

Figure 41.1 Outline of the sewage treatment process.

Primary Sedimentation

The degritted sewage passes into large settling tanks where the flow rate is further reduced allowing the less dense coarser organic particles to settle. The effluent passes on to secondary biological digestion and the settled sludge to sludge digestion.

Secondary Biological Treatment

Aerobic biological treatment is used to promote growth of a mixed culture of micro-organisms to break down the easily decomposible organic matter and to flocculate the finer particulates that clump by adhering to the cell surfaces. The most common form is the *activated sludge process* (Figure 41.2). The effluent from primary sedimentation is mixed with settled sludge from secondary sedimentation. This is mainly microbial cells and acts as an inoculum to promote rapid growth. Large aerators are used to maintain aerobic conditions.

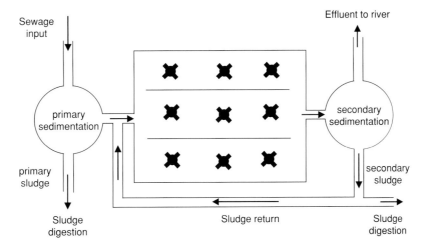

Figure 41.2 Schematic of the activated sludge process.

Secondary Sedimentation

After aerobic digestion the sewage is again allowed to settle. At this point the supernatant is often suitable for discharge into a river. The settled sludge passes to sludge digestion, with some being returned to the front of the secondary treatment ponds as activated sludge.

Tertiary Treatment

Where the effluent from secondary settling is being discharged into a particularly vulnerable environment, further treatment may be required. Phosphate stripping can be achieved by precipitation by adding lime or an iron salt, and ammonia stripping can be carried out by adding an excess of lime to raise the pH or by alternating aerobic and anaerobic digestion to promote nitrification of ammonium and nitrate removal by denitrification. Nutrients can be removed by reed bed systems, bacteria removed by disinfection using ultraviolet light or chlorine, and organic matter by filtration or adsorption on charcoal beds.

42. Environmental Impacts of Sewage

Effluent Disposal

Following secondary settling (or tertiary treatment) the effluent is discharged into a river. The 30:20 standard (30 mg/L suspended solids and 20 mg/L BOD_5) is based on a dilution by the river of 8:1. The natural decomposition of residual organic matter that takes place in the river is the final step of sewage treatment.

Figure 42.1 shows the impact of an unsatisfactory effluent, either because of inadequate treatment or insufficient dilution, on a receiving river. Upstream of the outfall the river has a low BOD_5, low suspended solids, is saturated with oxygen and is inhabited by a wide diversity of animal life. The effluent contains suspended solids, decomposable organic matter in solution and suspension (BOD_5) and plant nutrients such as ammonium, nitrite, nitrate and phosphate. The major impact is from the high BOD. Rapid decomposition by the heterotrophic micro-organisms in the river reduces the oxygen concentration in the decomposition and septic zones. In severe cases, anaerobic conditions may develop. Solids sediment to the river bed resulting in anaerobic conditions, and numbers of heterotrophic micro-organisms increase in response to the supply of nutrients. Further downstream the combined effects of decomposition and sedimentation have lowered the BOD and reduced the demand for oxygen so that oxygen levels start to recover, eventually returning to saturation several miles downstream of the input. This fall and recovery of oxygen saturation is known as the *oxygen sag curve*. In the decomposition processes ammonium and phosphate are released, and once the oxygen concentration starts to recover ammonium is oxidised to nitrate by nitrifying bacteria. Downstream, the plant nutrient levels will have increased, potentially leading to more prolific growth of rooted plants and phytoplankton.

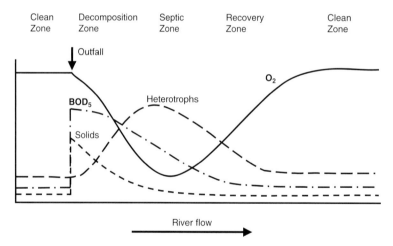

Figure 42.1 Impact of a poor quality sewage treatment works effluent on a river system.

The changes in oxygen saturation have profound effects on the river ecology. The clean water supports a diverse population of pollution intolerant invertebrates and fish (mayfly and stonefly nymphs, caddisfly larvae, trout, perch and bass). These organisms are absent from the oxygen depleted zones, where there is much less diversity of organisms. Worms, leeches, bloodworms and rat-tailed maggots are tolerant of the low oxygen concentrations. Bloodworms have high levels of haemoglobin which helps scavenge oxygen when it is at low concentration in the water, while the rat-tailed maggot uses a snorkel to obtain atmospheric oxygen. As the oxygen levels improve downstream, the pollution intolerant organisms return until eventually the full clean water population is restored.

Sludge Digestion

The combined primary and secondary sludges (a thicker dirty brown liquid at 5% total solids) undergo anaerobic digestion to reduce the pathogen load and stabilise the highly decomposable organic matter. This is a batch process taking 2–3 weeks, during which time up to 30% of the total solids is digested to biogas (a mixture of 70% methane and 30% carbon dioxide) greatly reducing the BOD and odour.

Sludge Utilisation/Disposal

The digested sludge can be considered as a useful manure containing N, P, K and organic matter, or as a problematic waste requiring disposal. Problems of high BOD, odour and residual pathogenic organisms remain; the potentially toxic elements present in sewage, largely as a result of industrial inputs, become concentrated in the sludge. Up until January 1999, dumping at sea was a major sink accounting for 30% of UK sludges. This is now banned under EU legislation. Disposal to farmland has always been, and remains, the largest disposal route, but public attitudes to food grown on sewage sludge treated soil and concern on the part of farmers about disease and the build up of potentially toxic elements mean that this option may become less available.

Disposal to non-agricultural land such as contaminated land sites is restricted by their limited area, and disposal to landfill is expected to decrease. Consequently, other disposal methods are becoming more important. Sewage sludge can be composted with straw or green waste to make it a more acceptable fertiliser treatment, but many of the concerns remain. Despite the high water content, it is possible to burn sludges after drying; this could include combustion with coal in power stations.

Eutrophication

Most water bodies have relatively low concentrations of the inorganic nutrients needed for plant growth (*oligotrophic* systems), and primary production by the phytoplankton is usually limited by phosphate or nitrate. *Eutrophic* water bodies have higher nutrient status and primary productivity (see Table 42.1). Eutrophication is a natural process whereby water bodies become enriched with organic matter and nutrients over time, but this can be greatly increased by pollution with nitrate and phosphate, principally from sewage treatment effluent and leaching from agricultural soils. About 70% of nitrate and 40% of phosphate in English waters are considered to be derived from agricultural sources; as greater efforts are made to reduce phosphate in sewage effluents, the agricultural contribution becomes more important. Eutrophication may result in the development of algal blooms and excessive growth of higher plants. Where nitrate and phosphate are both elevated, the resultant bloom is dominated by green algae and higher plants, while if phosphate is elevated but nitrate is limiting the bloom is composed mainly of blue-green algae (cyanobacteria), as many have the ability to fix atmospheric nitrogen.

Table 42.1 Nutrient level guidelines for assessing the trophic status of lake systems.

	Total P (μg/L)	Total N (μg/L)
Oligotrophic	<10	<200
Mesotrophic	10–20	200–500
Eutrophic	>20	>500

In addition to being unsightly, algal blooms may damage the aquatic ecosystem. Lower dissolved oxygen concentrations, the production of toxins (some cyanobacteria produce potent neurotoxins) and blockage of waterways by heavy plant growth can affect aquatic animals, such as invertebrates and fish and shellfish in marine waters, reduce the amenity and recreational value of the water, impact on navigation and impact on water use for drinking water or farm livestock to the extent that it may be necessary to restrict access to lakes.

43. Contaminated Land

In the UK, contaminated land issues are governed by Part IIA of the 1990 Environmental Protection Act. This legislation defines contaminated land as:

'Any land which appears to be in such a condition, by reason of substances in, on or under the land that significant harm is being caused, or there is a significant possibility of such harm being caused; or pollution of controlled waters is being, or is likely to be, caused.'

Harm is defined as: 'harm to the health of living organisms or other interference with the ecological systems of which they form a part, and in the case of man includes harm to his property'. *Substance* is defined as: 'any natural or artificial substance whether in solid or liquid form or in the form of gas or vapour'.

Another commonly used definition, formulated by a NATO committee, is:

'Land which contains substances that, when present in sufficient quantities, are likely to cause harm directly or indirectly to humans, the environment or other targets such as construction materials.'

Both of these definitions contain the concept of risk, which is encompassed in the concept of *pollutant linkage* (Figure 43.1).

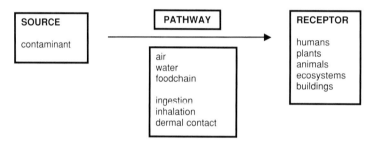

Figure 43.1 Pollutant linkage model.

This model embodies the idea that there is a source of a pollutant, which may have to exceed a certain critical concentration to be considered as a risk, and that there is a receptor, or range of receptors, that could be damaged by it. Crucially, there has to be a pathway via which the pollutant can reach the receptor. In relation to human health, the pollutant could be ingested, either directly as dust or indirectly by eating contaminated plants, inhaled as a gas or dust, or enter the body through the skin by direct dermal contact. In terms of damage to the environment or buildings, the pollutant can be transported either in air or water.

Extent of the Contaminated Land Problem

Estimates of the extent of the problem of contaminated land are difficult to make because different definitions have been used. Some contamination is obvious, for example waste from metal mining or old gas works sites, but diffuse pollution from industrial processes and historic contamination due to former industrial activity may not always be recognised.

In the UK, the Environment Agency has estimated that there may be up to 100 000 sites, covering up to 300 000 hectares, that could be defined as contaminated in England and Wales; in Scotland there are about 12 000 ha of derelict land (defined slightly differently), most of which is also contaminated. About 30% of contaminated land is affected by mining and associated activities, and 20% by general industrial contamination.

In the USA, the Environmental Protection Agency (USEPA) estimates that there are over 450 000 brownfield sites – defined as:

'a property, the expansion, redevelopment, or reuse of which may be complicated by the presence or potential presence of a hazardous substance, pollutant, or contaminant'.

In 1980 the Superfund Program was set up to deal with the worst sites, both to provide a rapid response where remedial action was required in an emergency, and to remediate the worst of abandoned sites. This is a rolling programme, which in 2005 had 1200–1300 sites on its National Priorities List.

Assessment of Risk on Contaminated Land

Because of the nature of contamination, and the requirement to show a pathway linking the source and receptor, a common protocol has developed which successively identifies potential contaminants and receptors that may be at risk, quantitatively measures the concentrations of contaminants, assesses the potential risk and then carries out remediation to remove or minimise that risk (Figure 43.2).

The key issue in this process is deciding what level of contamination poses a risk. In order to assess this, many countries have introduced a system of guidelines, which can be used to trigger action or intervention on a specific site. The aim is to provide non-specialists with guidance on whether or not remediation is required. The drawback of such schemes is that they usually consider only the total amount of contaminant, not its bioavailability, and they tend to be applied regardless of the chemical and physical nature of the contaminant. In the UK, the contaminated land exposure assessment (CLEA) model has recently been introduced, which models the probability of human exposure to contaminants in soil and is used to define soil guideline values (SGVs) for individual contaminants and specific land uses. In 2005, SGVs were available for As, Cd, Cr, inorganic Hg, Ni, Se, Pb, toluene and ethylbenzene (http://www.environment-agency.gov.uk/subjects/landquality/).

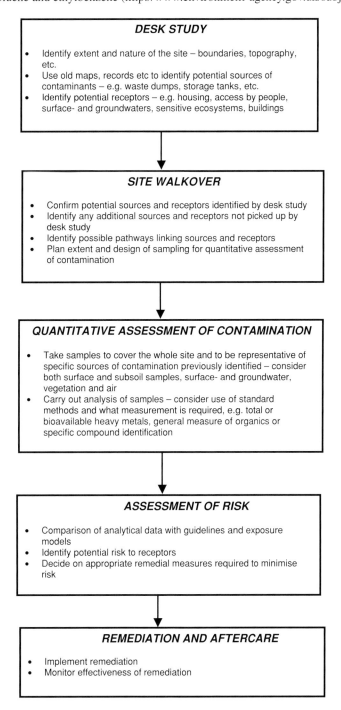

Figure 43.2 Stages in the assessment and remediation of contaminated land.

44. Organic Contaminants in the Environment

Many organic compounds are found as contaminants in the natural environment, mainly as a result of industrial processes or the burning of fossil fuels and other organic materials (pesticides are a specific example of such compounds and have been discussed in Topic 37). They can exist as liquids, vapours or particulates.

The main concern about organic contaminants is their persistence and their ability to accumulate and concentrate in biological tissue, mainly fatty tissue, and hence their long-term effects. Many are poorly soluble in water, but 1-octanol ($CH_3(CH_2)_6CH_2OH$) has been shown to be a suitable solvent for estimating their ability to accumulate in fat. The 1-octanol/water partition coefficient K_{ow} is used to predict the ability of an organic compound to bioaccumulate:

$$K_{ow} = [conc]_{octanol}/[conc]_{water}$$

Two broad categories of liquid organic contaminants are often recognised:

- *Light non-aqueous phase liquids (LNAPL)*, which include materials such as petrol, diesel and heating oil. These are less dense than water and so form a separate phase on top of water
- *Dense non-aqueous phase liquids (DNAPL)*, which include materials such as coal tar, creosote and chlorinated solvents, e.g. trichloroethene (TCE) and tetrachloroethene (PCE). These are more dense than water and so can sink through soil water to the groundwater and can form plumes within soils and rocks.

Much of the particulate material in air comes from burning of organic materials, especially fossil fuels, and so organics are a major component. The chemical make-up of particulate material is highly heterogenous. Particles are classified based on their size using the particulate matter (PM) scale:

PM_{10} – inhalable particles, which can enter the respiratory system
$PM_{2.5}$ – respirable particles, which can enter the lungs
$PM_{0.1}$ – ultrafine particles

The liquid and particulate organic materials contain many organic compounds that give rise to concerns about human and environmental health. Some of these are described below.

BTEX

The term BTEX is used to cover a group of simple aromatic compounds that are widespread within the environment: benzene, toluene, ethylbenzene and the xylenes (Figure 44.1). They are examples of LNAPLs. They are common solvents and also found in petrol. Benzene is a known human carcinogen.

Polychlorinated Biphenyls (PCBs)

This group of compounds are inert liquids with the general formula shown in Figure 44.2. There are 209 possible congeners, depending on the number and position on the ring of the chlorine atoms. The main use for PCBs was as coolants in power transformers and electrical equipment because of their good insulation properties and low flammability. They were also used as plasticisers, de-inking solvents and waterproofing agents. Production of PCBs in the USA was stopped in 1977, but they are still found in widespread use and are major environmental contaminants. They are poorly soluble in water, but are strongly adsorbed onto soil and sediment particles and are volatile, both of which provide a route for their dispersal. They are highly fat soluble and so bioaccumulate in animal tissues. PCBs have been found in the milk of marine mammals, such as seals in

Topic 44 Organic contaminants in the environment

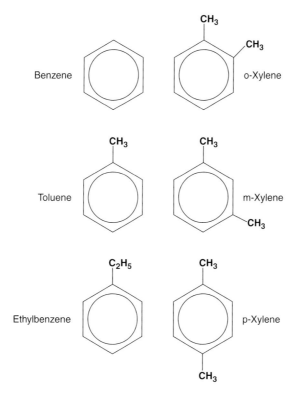

Figure 44.1 The BTEX compounds.

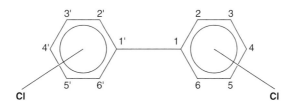

Figure 44.2 Polychlorinated biphenyls (PCBs).

polar regions, far removed from sources of pollution. They show an increase in concentration up a food chain that typifies bioaccumulation: e.g. phytoplankton, 0.0025 mg/kg; zooplankton, 0.1 mg/kg; fish, 1–5 mg/kg; eggs of fish eating birds, 100 mg/kg. There is little evidence of acute toxicity due to PCBs, but long term effects may relate to cancers and reproductive abnormalities. A common symptom to exposure is chloracne. Most human exposure to PCBs is by intake in food, although occasional accidental cases of PCB poisoning do occur, e.g. due to contaminated cooking oil in Japan (1968) and Taiwan (1979), and contaminated animal feed in Belgium (1999).

Dioxins and Furans

The terms 'dioxin' and 'furan' are commonly used to refer to two groups of compounds: *polychlorinated dibenzodioxins* (PCCDs, Figure 44.3) and *polychlorinated dibenzofurans* (PCDFs, Figure 44.4). There are 75 congeners of PCDDs and 135 congeners of PCDFs, depending on the number and positioning of chlorine atoms on the two rings.

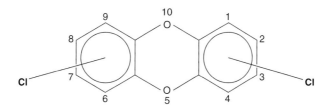

Figure 44.3 Polychlorinated dibenzodioxins (PCCDs).

107

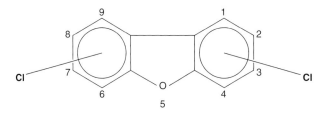

Figure 44.4 Polychlorinated dibenzofurans (PCDFs).

These substances are produced by fires (natural and artificial) and are also by-products in certain industrial processes. Two examples of dioxin contamination of the environment came about as a result of its presence as a contaminant in trichlorophenol. 2,3,7,8-Tetrachlorodibenzo-*p*-dioxin (2,3,7,8-TCDD) was a contaminant in the herbicide 2,4,5-T (2,4,5-trichlorophenoxyacetic acid), which was widely used as a defoliating agent (Agent Orange) during the Vietnam War in the 1960s; soils in this area still contain traces of dioxins, and there are concerns about health effects on Vietnamese citizens and American military personnel. In 1976 an explosion occurred at a factory in Seveso, Italy, which was manufacturing 2,4,5-trichlorophenol; as a result, dioxins were dispersed widely in the local environment causing the death of wildlife. Although there were no obvious short-term health effects on humans, there is now concern about increased numbers of some cancers. Pentachlorophenol is commonly used as a wood preserver, and the burning of treated wood produces octachlorodibenzo-*p*-dioxin. Bleaching of paper pulp with chlorine leads to the formation of 2,3,7,8-TCDD and the tetrachlorodibenzofurans 1,2,7,8-TCDF and 2,3,7,8-TCDF. This practice is now banned and alternative treatments such as ClO_2, hydrogen peroxide or ozone are used (the latter two producing 'chlorine-free' paper). There is little evidence of acute toxicity due to dioxins and furans, but long term effects may relate to cancers and reproductive abnormalities. A common symptom to exposure is chloracne.

Polybrominated Diphenyl Ethers (PBDEs)

This group of compounds has the general formula shown in Figure 44.5. There are 209 possible congeners, depending on the number and position on the ring of the bromine atoms. They are widely used as fire retardants in furniture, textiles, plastics and electronic equipment. They have only recently been recognised as pollutants. Like PCBs, they are fat soluble and bioaccumulate, and have been detected in animal tissues even in areas remote from a likely source. So far there are few data on their effects on human health, but it is thought that they may affect the neurosystem, hormone levels and the liver.

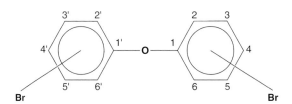

Figure 44.5 Polybrominated diphenyl ethers (PBDEs).

Polycyclic (or Polynuclear) Aromatic Hydrocarbons (PAHs)

This is a group of compounds that consist of two or more fused aromatic rings (i.e. adjacent rings share carbon atoms), those of environmental concern having up to six rings. Figure 44.6 shows three examples of PAHs.

The main source of PAHs is the combustion of organic material – wood burning, coal burning, petrol and diesel use, tobacco and charred food. They are major air pollutants, with those containing four rings or fewer existing in gaseous form, while those with more than four rings readily adsorb to soot and ash particles. PAHs are also found as water pollutants as the result of oil and petrol spills. The main health concern about PAHs is their carcinogenic property. Benzo[a]pyrene (Figure 44.6) is a potent carcinogen and is easily bioaccumulated (its K_{ow} is comparable to the organochlorine pesticides: log $K_{ow} = 6.3$).

Because of the large number of PAH compounds that exist, analysing for them in environmental samples is difficult. The United States Environmental Protection Agency (USEPA) uses the measurement of 16 of the most common PAHs and

Topic 44 Organic contaminants in the environment

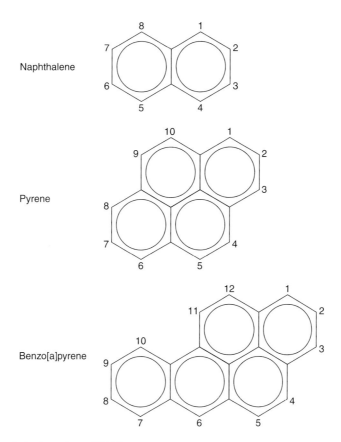

Figure 44.6 Polycyclic aromatic hydrocarbons (PAHs).

expresses PAH contamination as the sum of the 16 individual measurements (16-PAH). Although all 16 may not be present in all cases, the compounds measured are: acenaphthene, acenaphthylene, anthracene, benz[a]anthracene, benzo[a]pyrene, benzo[b]fluoranthene, benzo[ghi]perylene, benzo[k]fluoranthene, chrysene, dibenz[a,h]anthracene, fluoranthene, fluorene, indeno[1,2,3-cd]pyrene, naphthalene, phenanthrene and pyrene.

45. Heavy Metals

The term 'heavy metal', although commonly used, is difficult to define unambiguously. One way is to use a property of the elemental metal – relative density > 5 g/cm^3 and metals with atomic number >20 have been used. Part of the problem is that some heavy metals (termed *micronutrients*) are required in small amounts by plants, animals and micro-organisms (e.g. Cu, Zn), but are toxic at high concentrations; some metals have no known biological function (e.g. Cd, Pb); and some are not metals (e.g. As, Se). A term that is being increasingly used to cover this group of elements is *potentially toxic element* (PTE). The metals which are most commonly accepted as part of this group are: As, Cd, Cr, Cu, Hg, Ni, Pb, Se and Zn.

Arsenic (As)

Occurrence — Lithosphere: mean in igneous rocks 2 mg/kg, shales up to 250 mg/kg
Soil: typical range 1–50 mg/kg, mean 5 mg/kg
Main mineral, arsenopyrite (FeAsS); more commonly as impurity in Au, Cu, Pb and Zn minerals; relatively high in areas of geothermal activity.

Chemistry — As(III) as the arsenite ion AsO_3^{3-}; As(V) as the arsenate ion AsO_4^{3-}
Methylated species monomethylarsenic acid (MMA) and dimethylarsenic acid (DMA)
Chemisorbed onto hydrous oxides and kaolinite.

Requirement — No known requirement by plants; may be required in trace amounts by humans.

Toxicity — Highly toxic; carcinogenic
Phytotoxicity: soil >20 mg/kg, plant tissue >5 mg/kg.

WHO drinking water guideline — 0.01 mg/L.

UK soil guideline value (SGV) — Residential: 20 mg/kg
Commercial: 500 mg/kg

Uses — Wood preservative [chromated copper arsenate (CCA)]; pesticides.

Environmental problems — Burning of CCA-preserved wood; smelting of Au, Cu, Pb and Zn ores; coal burning; As containing pesticides.

Cadmium (Cd)

Occurrence — Lithosphere: mean 0.2 mg/kg
Soil: typical range 0.01–1 mg/kg, mean 0.05 mg/kg
Impurity substituting for Zn^{2+} in minerals.

Chemistry — Cd(II) as Cd^{2+} ion
Relatively soluble and bioavailable in environmental systems
Held by ion exchange; chemisorbed onto hydrous oxides; complexed by organic matter; precipitated as carbonate.

Requirement — No known requirement by plants or animals.

Toxicity — Highly toxic: long term effects on kidney function and bone metabolism ('itai-itai' disease), carcinogenic
Phytotoxicity: soil >3 mg/kg, plant tissue >5 mg/kg.

WHO drinking water guideline — 0.003 mg/L.

UK soil guideline value (SGV) — Residential: with plant uptake 2 mg/kg (at pH 7)
without plant uptake 30 mg/kg
Commercial: 1400 mg/kg.

Uses — Nickel–cadmium batteries; pigments; metal alloys, electroplating.

Environmental problems — Impurity in phosphate fertilisers added to agricultural soils; sewage sludge disposal; release into the atmosphere by fossil fuel burning and waste incineration.

Chromium (Cr)

Occurrence — Lithosphere: mean 200 mg/kg; <10 mg/kg in acidic igneous rocks, 10–100 mg/kg in sedimentary rocks, 100–200 mg/kg in basic igneous rocks, 2000–3000 mg/kg in ultramafic rocks (serpentine)
Soil: typical range 1–1000 mg/kg (depending on rock type), mean 100 mg/kg
Main mineral, chromite ($FeCr_2O_4$); commonly as impurity substituting for Al^{3+} in aluminosilicate minerals.

Chemistry — Cr(III) as chromic ion Cr^{3+}, highly stable, insoluble above pH 3.5, may complex with organic matter
Cr(VI) as chromate ion CrO_4^{2-} and dichromate ion $Cr_2O_7^{2-}$
Redox chemistry Cr(III) $\longleftrightarrow$ Cr(VI) important: soil organic matter may reduce Cr(VI) to Cr(III), while manganese oxides may oxidise Cr(III) to Cr(VI).

Requirement — Cr(III) required by animals for glucose metabolism; no known requirement by plants.

Toxicity — Cr(VI) is a known carcinogen
Phytotoxicity: soil >75 mg/kg, plant tissue >5 mg/kg (both values higher for serpentine soils where plants are adapted to tolerate high Cr).

WHO drinking water guideline — 0.05 mg/L.

UK soil guideline value (SGV) — Residential: with plant uptake 130 mg/kg
without plant uptake 200 mg/kg
Commercial: 5000 mg/kg.

Uses — Wood preservative [chromated copper arsenate (CCA)]; stainless steel production; tanning of leather; chrome plating; refractory use for high temperature furnace linings; magnetic recording tape; pigments; catalysts.

Environmental problems — Disposal of chromite ore processing residue (COPR); burning of CCA-preserved wood; disposal of sewage sludge to land.

Copper (Cu)

Occurrence — Lithosphere: mean 70 mg/kg
Soil: typical range 2–100 mg/kg, mean 30 mg/kg
Main mineral, chalcopyrite ($CuFeS_2$); commonly found as impurity in other sulfide ores; also found as oxide (cuprite Cu_2O, tenorite CuO), carbonate (malachite $Cu_2(OH)_2(CO_3)_2$) and as native metallic copper.

Chemistry — Cu(II) as Cu^{2+} ion
Forms very stable complexes with humified organic matter.

Requirement — Essential micronutrient for plants and animals. Required for functioning of some redox enzymes (oxidases); component of plastocyanin, a photosynthesis protein.

Toxicity — Relatively few problems of human toxicity
Phytotoxicity: soil >60 mg/kg, plant tissue >20 mg/kg.

WHO drinking water guideline — 2 mg/L.

Uses — Electrical and electronic equipment; plumbing and heating systems; pesticides; coins; wood preservative.

Environmental problems — Metal smelter emissions; sewage sludge disposal; disposal of pig and poultry manures; burning of CCA-preserved wood; pesticides.

Mercury (Hg)

Occurrence — Lithosphere: mean 0.1 mg/kg
Soil: typical range 0.02–0.2 mg/kg, mean 0.03 mg/kg
Main mineral, cinnabar (HgS)

Chemistry — Hg(I) as mercurous Hg^+ ion; Hg(II) as mercuric ion Hg^{2+}; methylated species methylmercury cation CH_3Hg^+.

Topic 45 Heavy metals

Requirement	No known requirement by plants or animals.
Toxicity	Highly toxic, especially organomercury compounds Phytotoxicity: soil >0.3 mg/kg, plant tissue >1 mg/kg.
WHO drinking water guideline	0.001 mg/L.
UK soil guideline value (SGV)	Residential: with plant uptake 8 mg/kg without plant uptake 15 mg/kg Commercial: 480 mg/kg.
Uses	Chloralkali industry; electrical equipment; dental amalgam; pigments; pesticides.
Environmental problems	Smelting of metal ores; fossil fuel burning.

Nickel (Ni)

Occurrence	Lithosphere: mean 100 mg/kg; <10 mg/kg in acidic igneous and sedimentary rocks, 10–100 mg/kg in sedimentary rocks, 100–200 mg/kg in basic igneous rocks, 2000–3000 mg/kg in ultramafic rocks (serpentine) Soil: typical range 5–500 mg/kg (mean 100 mg/kg) in soil derived from non-ultramafic rock, 500–10 000 mg/kg in soil derived from ultramafic rocks Main mineral pentlandite ($(Ni,Fe)_9S_8$), commonly as impurity in other sulfide minerals.
Chemistry	Ni(II) as Ni^{2+} ion Chemisorbed onto hydrous oxides, complexed by organic matter.
Requirement	Essential for some bacteria (in hydrogenase enzymes) and plants (uresae enzyme activity); may have a role in animal nutrition (unproven).
Toxicity	Highly toxic and carcinogenic Phytotoxicity: soil >100 mg/kg, plant tissue >10 mg/kg.
WHO drinking water guideline	0.02 mg/L.
UK soil guideline value (SGV)	Residential: with plant uptake 50 mg/kg without plant uptake 75 mg/kg Commercial: 5000 mg/kg.
Uses	Stainless steel production; alloys; Ni–Cd batteries; coins; electroplating; catalysts.
Environmental problems	Sewage sludge disposal; metal smelting.

Lead (Pb)

Occurrence	Lithosphere: mean 16 mg/kg Soil: typical range 2–200 mg/kg, mean 10 mg/kg Main mineral galena (PbS), also cerrusite ($PbCO_3$) and anglesite ($PbSO_4$).
Chemistry	Pb(II) as Pb^{2+} ion Forms very stable complexes with humified organic matter, precipitated by phosphate, chemisorbed by hydrous oxides.
Requirement	No known requirement by plants or animals.
Toxicity	Highly toxic Phytotoxicity: soil >100 mg/kg, plant tissue >30 mg/kg.
WHO drinking water guideline	0.01 mg/L.
UK soil guideline value (SGV)	Residential: 450 mg/kg Commercial: 750 mg/kg.
Uses	Batteries; radiation shielding; (historic, but decreasing uses – pesticides, paints, anti-knock agent in petrol, plumbing).
Environmental problems	Metal smelting; sewage sludge disposal; legacy of petrol derived lead.

Selenium (Se)

Occurrence	Lithosphere: mean 0.1 mg/kg Soil: typical range 0.1–2 mg/kg, mean 0.3 mg/kg.
Chemistry	Se(−II) as selenide Se^{2-} ions in reduced environments; Se(IV) as selenite SeO_3^{2-} ion; Se(VI) as selenate SeO_4^{2-} ion; organoselenium compounds, e.g. dimethylselenide $(CH_3)_2Se$ Selenite and selenate chemisorbed by hydrous oxides.
Requirement	Essential micronutrient for animals; antioxidant, interacts with vitamin E metabolism; has a role in some enzyme systems.
Toxicity	Highly toxic Phytotoxicity: soil >5 mg/kg, plant tissue 5 mg/kg.
WHO drinking water guideline	0.01 mg/L.
UK soil guideline value (SGV)	Residential: with plant uptake 35 mg/kg without plant uptake 260 mg/kg Commercial: 8000 mg/kg.
Uses	Dietary supplement; some shampoos; glass making; pigments; agrochemicals; antioxidant.
Environmental problems	Burning of fossil fuels; sewage sludge disposal.

Zinc (Zn)

Occurrence	Lithosphere: mean 80 mg/kg Soil: typical range 10–300 mg/kg, mean 50 mg/kg.
Chemistry	Zn(II) as Zn^{2+} ion Relatively soluble and bioavailable in environmental systems Held by ion exchange; chemisorbed onto hydrous oxides; complexed by organic matter.
Requirement	Essential micronutrient for plants and animals, mainly for functioning of enzymes (over 20 Zn enzymes identified).
Toxicity	Relatively few problems of human toxicity Phytotoxicity: soil >70 mg/kg, plant tissue >100 mg/kg.
WHO drinking water guideline	No value.
UK soil guideline value (SGV)	No value.
Uses	Galvanising of steel; alloys; pigments; some shampoos.
Environmental problems	Metal smelting; burning of fossil fuels; sewage sludge disposal.

46. Environmental Impacts of Mining

Mining of economically important resources is a major global business (Table 46.1) which can have a number of environmental consequences. Broadly there are two techniques used for mining: *surface* or *opencast mining* and *deep mining*. In opencast mining the rock strata and soil are carefully stored during the mining operation and are then replaced to try and reform a landscape similar to the original pre-mining one. This process presents different problems to that of deep mining, mainly ones of landscape, drainage and vegetation establishment. Deep mining involves removal of rock during the sinking of a shaft and removal of rock strata from around the material being mined. This material is then dumped in a tip on the surface.

Table 46.1 Global mineral production figures for 2003.

Mineral	Production in 2003
Coal	2.7 Gt
Iron ore	1.0 Gt
Lime	0.117 Gt
Potash	0.028 Gt
Aluminium	0.022 Gt
Kaolin	0.021 Gt
Chromite	0.014 Gt
Copper	0.014 Gt
Manganese	0.012 Gt
Zinc	0.009 Gt
Titanium	0.005 Gt
Lead	0.003 Gt
Asbestos	0.002 Gt
Tin	220 Kt
Antimony	155 Kt
Gold	2.3 Kt
Nickel	1.3 Kt
Silver	0.9 Kt

Problems Due to Mine Spoil Tips

Size and Shape of Tip – Stability, Visual Impact

The waste is inherently unstable, consisting of various sizes of rock fragments with no binding mechanism to hold them together. The waste forms tips with steep slopes at natural angle of repose of about 37°. This can lead to *slump*, the falling away of large sections of a slope; *erosion*, especially due to water flow once weathering has resulted in build-up of sufficient fine material to decrease infiltration rate, causing loss of surface material and gully formation; or *creep*, a more gradual process due to gravity and possibly water, which can cause burial of naturally established plants. Flat areas around tips and on plateaux tend to have a well-weathered surface, which leads to impermeability (roots and water) and poor water-holding.

Physical Properties of Final Surface/Growth Medium – Root Penetration, Waterholding Ability, Erosion

Physical problems on reclaimed spoil tips are due to poor texture (coarse sandy material or weathering to fine particles), lack of humified organic matter (poor structure) and to reclamation practices (compaction). These cause poor water infiltration and drainage (waterlogged in wet periods, drought in dry periods) and formation of a surface pan (poor root penetration, anaerobic conditions). Treatments to improve physical conditions include: ripping or deep cultivation to create drainage pores; installation of drainage system, usually ditches; incorporation of organic matter; establishment of vegetation; spreading of topsoil.

Nutrient Status of Growing Medium

Because mining spoil is essentially crushed rock with little or no humified organic matter, nitrogen is the major nutrient that is deficient and N supply is the main problem to be addressed. Important aspects of this are: introduction of N into spoil by means of fertilisers, manures or vegetation (legumes); breakdown and incorporation of N into spoil by biological activity to establish a pool of N and N cycling.

Mine Tailings

A second type of problem from both opencast and deep mining arises from the treatment of the mined material. This usually involves some or all of the following processes, crushing, washing, sedimentation and separation, that result in a fine slurry of mineral particles called *tailings*, which are usually disposed of in storage lagoons. Depending on the material being mined, tailings may be acidic, may contain high concentrations of heavy metals from the mined material or of chemicals used to treat the mined material. Environmental problems occur when the walls of tailings lagoons are breached. Two recent examples in Europe were from the Aznalcollar mine in Spain, where pyrite and heavy metal sulfides were mined, and the Baia Mare mine in Romania, where cyanide was used to extract gold and silver from tailings from previous mining operations. At Aznalcollar about 5 million cubic metres of waste were released, which was highly acidic ($\sim$pH 3) and contained high concentrations of heavy metals, especially As, Cu, Pb and Zn (0.1–1%). The waste flowed into the Agrio and Guadiamar Rivers and spilled over to affect farmland up 0.5 km on each side of the river for a distance of about 40 km, affecting agricultural crops and the ecology of the area. At Baia Mare 0.1 million cubic metres of waste containing cyanide and high amounts of heavy metals were released into the rivers Somes, Tisza and Danube, causing large numbers of fish deaths for over 600 km.

Acid Mine Drainage

The major environmental problem in some, but not all, metal and coal mining spoils is the production of acid due to the oxidation of pyrite (FeS_2) and other metal sulfides. When pyrite is exposed to air and water in the spoil tip, it oxidises to produce sulfuric acid. This causes very acidic conditions in the waste and, when leached out of the waste in streams and rivers. There are two reaction pathways by which pyrite can oxidise:

> Oxidation by oxygen
> $$4FeS_2 + 15O_2 + 14H_2O \longrightarrow 4Fe(OH)_3 + 8H_2SO_4$$
>
> Oxidation by Fe(III)
> $$FeS_2 + 14Fe^{3+} + 8H_2O \longrightarrow 15Fe^{2+} + 2SO_4^{2-} + 16H^+$$

Oxidation of pyrite by Fe^{3+} ions is catalysed by the bacterium *Thiobacillus ferrooxidans*, which is active only below approximately pH 4. This catalyses the oxidation of Fe^{2+} to Fe^{3+} and proceeds at a rate that is 10^6 times faster than oxidation by O_2. Thus once pH falls to below 4 there is a massive increase in the rate of acid production.

Acid mine drainage from pyritic spoil (and from abandoned mines) has a low pH (typically less than 2) and is high in Al, Fe, Mn and other heavy metals (Zn, etc.). This causes problems in receiving watercourses. The acidity has a direct effect on living organisms, and the high concentration of soluble metals (Al, etc.) is toxic. The dissolved Fe^{3+} ions are insoluble at the higher pH of the receiving watercourses and so precipitate as iron oxide (ochre), coating the streambed and affecting the stream ecology.

Acidity is not the same as pH; it is made up of *actual acidity* (H^+) and *potential acidity* due to hydrolysis of Fe^{3+}, Al^{3+}, Mn^{2+}, etc. So *acidity* is defined as the capacity of a solution to neutralise a strong base (0.1 M NaOH) to a specified end-point (pH 8.3; phenolphthalein). *Alkalinity* is defined as the capacity of a solution to neutralise a strong acid (0.8 M H_2SO_4) to a specified end-point (pH 4.5; bromocresol green–methyl red). Most mine waters have some alkalinity due to the presence of bicarbonate (HCO_3^-) ions.

> Net acidity = (total acidity) − (total alkalinity)

Acidity of mine drainage (as mg/L $CaCO_3$ equivalent) can also be calculated by measuring the pH and the concentrations of the hydrolysable cations:

> Calculated acidity $= 50[2Fe^{2+}/56 + 3Fe^{3+}/56 + 3Al/27 + 2Mn/55 + 1000(10^{-pH})]$

It is estimated that there are 4500 miles of streams and rivers in the mid-Atlantic area of the USA (Delaware, District of Columbia, Maryland, Pennsylvania, Virginia and West Virginia) affected by acid mine drainage.

Acid mine drainage can be actively treated by addition of alkaline chemicals such as limestone ($CaCO_3$), hydrated lime [$Ca(OH)_2$], soda ash (Na_2CO_3), caustic soda (NaOH) or ammonia solution (NH_4OH). This is a widely used technology, which gives good control of pH and low residual metal concentrations, but is expensive and generates large amounts of sludge. Passive treatment of acid mine drainage involves systems in which it flows through a well containing crushed limestone, or through open ditches filled with limestone, or anoxic limestone drains (closed ditches filled with limestone). The limestone adds alkalinity, which raises pH and precipitates the metals. There is, however, a problem of iron oxide coating the limestone, thus decreasing its effectiveness. More recently, use has been made of constructed wetlands, which use soil- and water-borne microbes associated with wetland plants to remove dissolved metals from mine drainage. They are relatively cheap, having low running costs and requiring little or no maintenance, but they are a relatively new technology that is not yet fully understood.

Metal Smelting and Coal Burning

The burning of coal or the smelting of metal ores can be a major source of atmospheric pollution due to acidity (see Topic 52) and the release of heavy metals (see Topic 45). The case of the Ni/Cu smelter at Sudbury in Ontario, Canada, is a good example of the environmental consequences of metal mining and smelting which occurred there for over 100 years, and is also a major success story of remediation of the devastated areas. The emissions from the smelter were highly acidic due to the production of SO_2 as a result of roasting the sulfide ores:

$$2NiS + 3O_2 \longrightarrow 2NiO + 2SO_2$$

The sulfur dioxide caused severe acidification of soils (pH < 3) and particulate material deposited heavy metals (especially As, Cu and Ni) over a wide area. As a result of these emissions, 17 000 ha of land was completely devoid of vegetation and over a further 64 000 ha of vegetation was stunted. In recent years, a successful revegetation programme has established plant cover, especially trees, over much of this area by the use of lime to correct the pH and fertiliser to supply nutrients.

47. Radioactivity

Atoms of the same element all have the same number of protons and therefore atomic number, but *isotopes* of the element can exist with different numbers of neutrons and therefore different relative atomic masses: e.g. $^{12}_{6}C$, $^{13}_{6}C$ and $^{14}_{6}C$ are all isotopes of carbon (atomic number 6). The superscript number refers to the relative atomic mass (the number of protons plus neutrons) and the subscript number to the atomic number (the number of protons). A *nuclide* is an atom whose nucleus contains a specified number of protons and neutrons. Nuclei with an unstable ratio of neutrons to protons (*radionuclides*) undergo spontaneous disintegration (*radioactive decay*) to form a more stable ratio by the emission of a charged particle and/or electromagnetic radiation; for example:

$$^{14}_{6}C \longrightarrow {}^{14}_{7}N + \beta^{-}$$
$$^{238}_{92}U \longrightarrow {}^{234}_{90}Th + {}^{4}_{2}He^{2+}$$

> The three main types of radioactive emission (see Table 47.1) are:
>
> *Alpha particle*   Heavy positively charged particle (helium nucleus $^{4}_{2}He^{2+}$)
> *Beta particle*   Very small negatively charged particle (electron)
> *Gamma radiation* Very short wavelength electromagnetic radiation

The unit of measurement is the becquerel (Bq) which represents one disintegration per second.

The Half-Life of Radioactive Decay

The *half life* is the time taken for the concentration of the radioactive substance to fall by half (Figure 47.1). This half life is characteristic of a particular radionuclide, it is dependant solely on the structure of the atom and is not affected by physical or chemical conditions. Half lives may vary from millions of years to fractions of a second, depending on the stability of the nucleus.

Table 47.1 Penetrative properties of α, β and γ radiation.

Type of emission	Ionisation intensity	Relative range	Distance moved in air	Penetration of solids
α	High	Short	2–10 cm	0.1 mm tissue
β	Medium	Medium	1–2 m	Few mm glass
γ	Low	Long	Many metres	5–10 cm lead

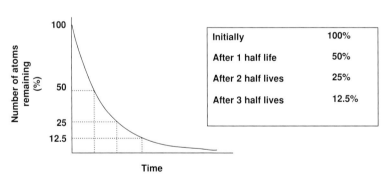

Initially	100%
After 1 half life	50%
After 2 half lives	25%
After 3 half lives	12.5%

Figure 47.1 Radioactive half life.

Primordial Radionuclides

Primordial radionuclides are those which, because of their very long half lives, remain from the matter that formed the Earth 4.6 billion years ago (Table 47.2). Radionuclides with shorter half lives have decayed to the point that they are no longer detectable.

Several of the heavy radionuclides e.g. $^{232}_{90}Th$, $^{238}_{92}U$ and $^{235}_{92}U$, are the starting point of a radioactive decay series (Figure 47.2), a series of daughter radioactive elements eventually leading to a final stable product. The daughter radionuclides can be detected, despite often very short half lives, because they are continually produced by decay of the parent.

Table 47.2 Some primordial radionuclides and their half lives.

Radionuclide		Half life (years)
$^{235}_{92}U$	Uranium-235	7.1×10^{8}
$^{40}_{19}K$	Potassium-40	1.3×10^{9}
$^{238}_{92}U$	Uranium-238	4.5×10^{9}
$^{232}_{90}Th$	Thorium-232	1.4×10^{10}
$^{87}_{37}Rb$	Rubidium-87	4.9×10^{10}

Cosmogenic Radionuclides

Cosmogenic radionuclides are produced when cosmic rays from space strike the nuclei of atoms such as C, N and O in the atmosphere. They have much shorter half lives than the primordial radionuclides but are being continuously produced, so an equilibrium abundance exists depending on the rate of production and rate of decay (Table 47.3).

Table 47.3 Some cosmogenic radionuclides and their half lives.

Radionuclide		Half life
$^{3}_{1}H$	Tritium	12.3 years
$^{14}_{6}C$	Carbon-14	5730 years
$^{7}_{4}Be$	Beryllium-7	53 days
$^{26}_{13}Al$	Aluminium-26	740 000 years

Artificial Radionuclides

Man has been experimenting with and using radionuclides for a little over a century, and has enhanced the natural background of radioactivity by the release of artificial radionuclides (Table 47.4). Nuclear fission power plants, nuclear fuel reprocessing, nuclear accidents, atomic weapons testing and medical uses have all contributed to the increase (Table 47.5).

Table 47.4 Some artificial radionuclides and their half lives.

Radionuclide		Half life
$^{3}_{1}H$	Tritium	12.3 years
$^{14}_{6}C$	Carbon-14	5730 years
$^{131}_{53}I$	Iodine-131	8.04 days
$^{129}_{53}I$	Iodine-129	1.57×10^{7} years
$^{137}_{55}Cs$	Caesium-137	30.2 years
$^{135}_{55}Cs$	Caesium-135	3.01×10^{6} years
$^{90}_{38}Sr$	Strontium-90	28.8 years
$^{239}_{94}Pu$	Plutonium-239	2.41×10^{4} years

Cosmic Radiation

Primary cosmic radiation is mainly very high energy photons; some of this radiation comes from our sun, but much comes from outside the solar system. Little of the primary cosmic radiation penetrates the atmosphere because collisions with atomic nuclei produce cosmogenic radionuclides and a shower of lower energy, secondary cosmic radiation (photons, electrons, neutrons and muons), some of which reaches the surface. The effect of the atmosphere on reducing cosmic radiation can be seen by the effect of altitude on exposure. Frequent fliers and air crew experience significantly increased exposure depending on the duration and altitude of their flights (Table 47.6).

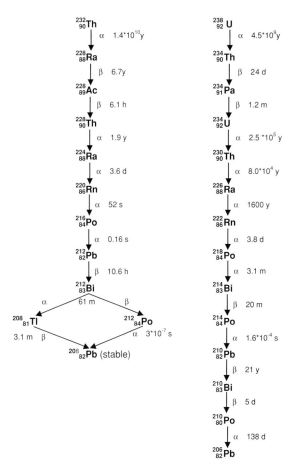

Figure 47.2 The thorium-232 and uranium-238 decay series (y = year; d = day; m = minute; s = second).

Table 47.5 Total global releases of artificial radioactivity.

	Total global release (TBq)*		
	^{239}Pu	^{137}Cs	^{14}C
Bomb testing	10 000	1 000 000	200 000
Reprocessing	600	50 000	200
Chernobyl	50	40 000	Small
Reactors	Small	Small	1000

*Units are terabecquerel; tera = 10^{12}.

Table 47.6 Effect of altitude on exposure to cosmic radiation and typical annual exposures.

a

Altitude (km)	Exposure (μSv/h)
15	10
10	5
5	1
2	0.1
Sea level	0.03

b

Group	Annual exposure (mSv)
Average UK population	0.26
Frequent flyer	0.7
Air crew	5.0

48. Environmental Impacts of Radioactivity

Biological Effects of Radioactivity

Alpha, β and γ radiation damages biological molecules such as proteins, enzymes and DNA by its ionising effects (Figure 48.1). This damage leads to symptoms of radiation sickness. The dose of received radioactivity, related to the energy absorbed by the tissue, is measured in sievert (Sv).

In cases of a single acute exposure to high levels of radioactivity, such as an accident, the severity of the symptoms increase with the radioactive dose. There is a threshold below which no symptoms are observed; at higher doses increasingly severe symptoms and finally death occur (Figure 48.2).

The likely consequence of chronic exposure to low levels of radioactivity, such as exposure to environmental radioactivity, is an increase in cancers. In this case the probability of an effect, rather than the severity, increases with exposure; there is probably no safe threshold (Figure 48.3).

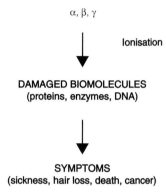

Figure 48.1 Effect of radioactivity on living organisms.

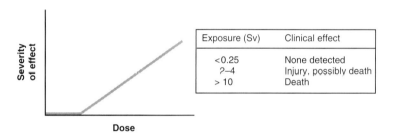

Figure 48.2 Effect of acute exposure to radioactivity.

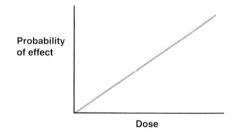

Figure 48.3 Effect of long term low level exposure to radioactivity.

Exposure to Radioactivity

The average exposure of the UK population to radioactivity is 2.6 millisievert per year (mSv/year). The largest component is from the naturally occurring radioactive gas radon and its decay products. Eighty-seven per cent of the exposure is from naturally occurring sources and only 13% from artificial radioactivity. However, 12% comes from the medical uses of radioactivity, leaving only 1% for other artificial sources, such as the nuclear industry and fallout from atmospheric atomic weapons testing (Table 48.1).

Variations in exposure across the UK relate primarily to geology (igneous rocks rich in ^{238}U are the main source of radon, and rock type influences the terrestrial γ radiation level) and to some extent to occupational exposure (Table 48.2).

Table 48.1 Average UK exposure to radioactivity by source.

Source of exposure	% of total
Radon and daughter nuclides	51
Terrestrial gamma radiation	14
Internal body burden	12
Cosmic radiation	10
Medical	12
Non-medical artificial	1

Table 48.2 Examples of variations in exposure to radioactivity in the UK.

	Exposure (mSv/year)
Average UK	2.6
Medical radiation workers	2.7
Nuclear industry workers	3.6
Frequent flyer	3.0
Air crew	7.6
Living in Cornwall	7.8

By European standards, exposure levels in the UK are low. This is due to generally higher levels of exposure to radon and environmental γ radiation in other countries (Figure 48.4).

Radon Exposure

Exposure to the radioactive gas radon accounts for over 50% of total exposure to radioactivity (Table 48.1). Radon is formed in the ^{232}Th and ^{238}U decay series (Figure 47.2), the main source being rocks containing ^{238}U, particularly in south-west England, Wales and northern Scotland (Figure 48.5).

Radon gas produced by the ^{238}U and ^{232}Th decay series escapes through fissures in the rock and in the open air quickly disperses into the atmosphere, but higher levels can accumulate inside buildings. Radon and its daughter radionuclides can be inhaled into the lungs where they decay producing α, β and γ radiation, thus increasing the risk of lung cancer. The main factors controlling indoor radon levels are uranium content and extent of fissuring in the rock, and the level of ventilation in the home. In a UK survey, 12% of homes in Devon and Cornwall had radionuclide levels exceeding 200 Bq/m^3, in contrast to an average UK concentration of 20 Bq/m^3.

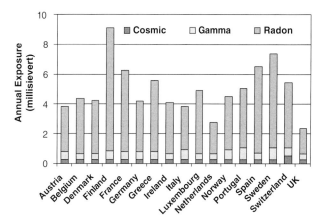

Figure 48.4 Average exposure to environmental radioactivity in European countries.

Radionuclide Dating Techniques

A range of dating techniques are used in environmental studies. The primordial radionuclides are used for dating rocks and determining the age of the Earth at the scale of hundreds of millions or billions of years. Radiocarbon dating is used for biological samples over tens of thousands of years and ^{210}Pb dating for measuring sediment accumulation over a timescale of a few hundred years.

Radiocarbon Dating

^{14}C is a cosmogenic radionuclide constantly being produced by cosmic rays striking the atmosphere. Consequently, atmospheric carbon dioxide is labelled with a predictable level of ^{14}C. Plants take up the ^{14}CO$_2$ during photosynthesis and their tissues become labelled. Animal and microbial tissues become labelled when they consume plant tissues. All living organisms have a ^{14}C content in equilibrium with the atmospheric level and the half life for ^{14}C decay, 5730 years (226 Bq/kg C). When the organism dies it stops taking in further ^{14}C and the ^{14}C present decays. Provided the sample is preserved (e.g. wood, cloth, bone, shells or charcoal from a fire), it is possible to calculate the time since the death of the organism by measuring the ^{12}C:^{14}C ratio. Because of fluctuations in the rate of production of ^{14}CO$_2$ in the atmosphere due to changes in the cosmic ray flux and more recently anthropogenic effects (CO$_2$ from burning of fossil fuels has no ^{14}C, inputs from atomic weapons testing) greater accuracy has been achieved by calibrating the method against other dating techniques.

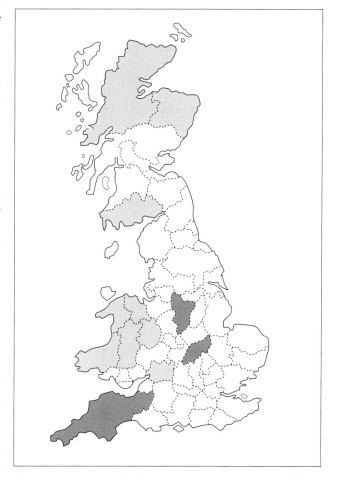

Figure 48.5 Areas with high and medium radon levels (■ high; ☐ medium).

^{210}Pb Dating

^{210}Pb is produced from the decay of radon gas in the atmosphere. This results in the deposition of ^{210}Pb from the atmosphere. In areas of rapid sediment or peat accumulation, the ^{210}Pb adheres to the sediment particles or is bound to the peat and is incorporated in the sediment. When first laid down, the sediment is uniformly labelled with ^{210}Pb which decays with a half life of 21 years. Provided the sediment is not disturbed by vertical mixing, the decline in ^{210}Pb down the profile can be used to calculate the age of the sediment at any depth and the rate of sediment accumulation.

49. Global Warming and Climate Change

Energy Balance at the Surface of the Earth

The atmosphere and surface of the Earth are warmed by incoming radiation from the sun, which consists primarily of wavelengths in the visible (0.40–0.75 μm) and infrared (IR) (0.75–3 μm) parts of the electromagnetic spectrum. Only very little ultraviolet (UV) radiation (<0.40 μm) reaches the Earth's surface, being largely absorbed in the stratosphere. Some of the longer wavelength radiation is also absorbed by the atmosphere, but overall about 50% of the incoming solar radiation reaches the surface of the Earth. The warm Earth radiates energy back into the atmosphere as IR radiation (Figure 49.1). Most of this radiated outgoing IR energy is absorbed by molecules in the atmosphere, especially CO_2, and then randomly re-emitted in all directions. Some is lost to space, but some returns to the Earth's surface causing further heating of the surface and the air in the lower atmosphere. This is the *greenhouse effect*, which is an entirely natural phenomenon

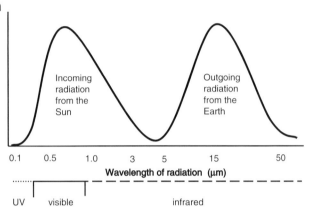

Figure 49.1 Wavelengths of incoming solar radiation and outgoing radiation from the Earth.

and is reposnsible for maintaining the Earth's surface temperature at an average of $+15°C$, so allowing life to exist. The current concern about global warming and climate change stems from the possible effects of enhanced CO_2 concentrations due to fossil fuel burning and an increase in other heat absorbing molecules.

Energy Absorption by Greenhouse Gases

Atoms in molecules move relative to one another in vibrations with a specific frequency. Two types of such movements occur; stretching vibrations and bending vibrations. In the first case, stretching vibrations, the atoms of the molecule move away from and towards each other causing the bond lengths to repeatedly lengthen and shorten. In bending vibrations, the angle between two adjacent bonds increases and decreases. Of the greenhouse gases, CO_2 shows stretching vibrations and H_2O shows bending vibrations.

If the energy of the outgoing radiation matches exactly the frequency of the molecular vibrations, then the molecule will absorb that radiation. In order to vibrate in this way a molecule must consist of more than one atom, so the inert gases, such as argon, do not absorb IR radiation. A second requirement is that the atoms in the molecule are different, otherwise there is no change in the electric field within the molecule, i.e. they have no dipole moment. Thus N_2 and O_2 do not absorb IR radiation.

The Greenhouse Gases

Carbon dioxide and water vapour are particularly effective at absorbing IR radiation of wavelengths below 8 μm and above 14 μm, leaving a window in the spectrum where most of the radiation passes through the atmosphere and is lost to space. Part of the concern about an enhanced greenhouse effect, which may lead to global warming, is due to increasing amounts of gaseous molecules in the atmosphere that absorb IR radiation within that part of the spectrum (Table 49.1). Another concern is the increase in CO_2 concentration, which has risen over the past 200 years from 280 ppm before the start of the Industrial Revolution to approximately 380 ppm at the start of the 21st century. Evidence from analysis of air bubbles trapped in the Antarctic ice sheet shows that such large fluctuations in CO_2 concentrations have occurred over the last 150 000 years, as have methane concentrations, and that these correlate well with temperature changes over that time. While this does not prove a cause and effect, it does seem that the changes in concentration of the gases and temperature are related.

Water Vapour

Although the amount of water vapour in the atmosphere is highly variable in both space and time, with an average concentration of 0.4%, it could be considered as the most important greenhouse gas. As there are no significant anthropogenic sources of water vapour contributing to the amount in the atmosphere, its effect is entirely natural and does not contribute to the

Topic 49 Global warming and climate change

Table 49.1 The greenhouse gases.

Gas	Current conc.	Approximate lifetime in atmosphere (years)	Greenhouse effect relative to CO_2	Absorbing range (μm)
CO_2	380 ppm	50–200	1	4.0–4.3 and 14–19
CH_4	1.8 ppm	10	23	3.0–3.6 and 7.1–8.3
N_2O	320 ppb	150	296	7.4–8.7
H_2O	Variable			2.5–3.0, 5–8 and >14
O_3	Variable			9–10
CFC 11	0.26 ppb	50	4600	8–12
CFC 12	0.53 ppb	100	10500	8–12
HCFC 22	0.15 ppb	10	1700	8–12
HFC 134a	0.01 ppb	15	1300	8–12
Halon 1301	0.003 ppb	65	7000	8–12
SF_6	0.005 ppb	3200	22200	8–12

enhanced greenhouse effect. Water vapour does, however, play an important role in controlling temperature fluctuations by its involvement in two feedback mechanisms.

Positive feedback	increased global warming ⟶ increased evaporation of water ⟶ more atmospheric water vapour ⟶ increased warming
Negative feedback	increased global warming ⟶ increased evaporation of water ⟶ more cloud cover ⟶ increased reflection and absorption of incoming solar radiation ⟶ less solar radiation reaches surface ⟶ decreased warming

Carbon Dioxide (CO_2)

Carbon dioxide is the second most important greenhouse gas after water vapour. It is produced naturally by respiration by aerobic organisms, decay of organic matter, natural fires and up-welling of ocean waters. There are also significant natural sinks for carbon: photosynthesis, and hence biomass carbon, dissolution in the oceans, carbonate formation and organic soils and sediments. These natural sources and sinks are roughly in balance (see Topic 31). Anthropogenic sources of CO_2, primarily fossil fuel burning and deforestation, release about 5 Gt C per year into the atmosphere.

Methane (CH_4)

Wetlands are the main natural source of methane, and along with production by ruminant animals, termites, wild fires, oceans and volcanoes, about 0.15–0.25 Gt CH_4 is released per year into the atmosphere. Anthropogenic sources, including rice paddies, natural gas useage, landfill sites, mining, biomass burning and agriculture, contribute a further 0.2–0.3 Gt CH_4 per year. Most of this methane, both natural and anthropogenic, is a result of anaerobic respiration by micro-organisms (see Topic 12). The only natural sink for methane is oxidation by hydroxyl radicals in the troposphere. Methane is second to carbon dioxide as a contributor to the enhanced greenhouse effect.

Nitrous Oxide (N_2O)

Nitrous oxide is produced naturally by microbial nitrification and denitrification in soils and waters, with about 0.007–0.014 Gt N_2O per year being emitted. Anthropogenic sources are the use of fertilisers and manures in cultivated soils (promoting nitrification or denitrification), biomass burning, fossil fuel burning, landfill, sewage disposal and nylon production. Together these add a further 0.005–0.006 Gt N_2O per year into the atmosphere. There are no known natural sinks for nitrous oxide.

Halogenated Gases (CFCs, Halons, SF_6)

These are wholly anthropogenically produced compounds. They were widely used in freezers, refrigerators and air conditioning systems. Their use was curtailed by the Montreal Protocol in 1987 (see Topic 51).

Estimates to assess the influence of the greenhouse gases to enhanced warming (i.e. over and above the natural greenhouse effect) suggest that carbon dioxide contributes about 55%, the halogenated gases 25%, methane 15% and nitrous oxide 5%; the effect of ozone is uncertain. There are international agreements in place to try to limit the emissions of greenhouse gases. This process started at the Earth Summit held in Rio de Janiero in 1992 and continued with the Kyoto Protocol in 1997, when 39 industrial countries agreed to decrease their greenhouse gas emissions. The Kyoto Protocol specifically limits emissions of carbon dioxide, methane, nitrous oxide, hydrofluorocarbons, perfluorocarbons and sulfur hexafluoride.

50. The Ozone Layer

The Ozone Layer

The ozone molecule (O_3) comprises three oxygen atoms and is formed naturally in the upper atmosphere, particularly in the lower stratosphere (the ozone layer), by the action of ultraviolet (UV) radiation from the sun on oxygen molecules. The concentration of ozone in the ozone layer is extremely low: if all the ozone were concentrated into a layer at sea level it would only be about 4 mm deep. However, the ozone layer is critical in preventing sufficient of the sun's ultraviolet radiation from reaching the Earth's surface to allow life on the planet.

The ozone layer depends on an equilibrium between ozone forming reactions and ozone destroying reactions.

Ozone is formed by the action of UV light splitting an oxygen molecule into two oxygen atoms, each of which can then react with a further oxygen molecule to form ozone. In the second reaction a neutral third molecule (M), usually N_2 or O_2, is needed to absorb the energy that is released in the reaction as heat. Note how the trend of cooling with increased altitude in the troposhere is reversed in the stratosphere (see Figure 50.1).

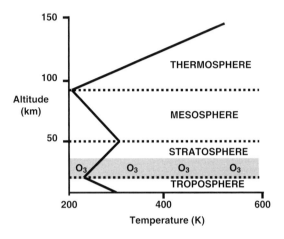

Figure 50.1 The ozone layer.

$$O_2 + h\nu(\lambda < 240 \text{ nm}) \longrightarrow 2O$$
$$O_2 + O + M \longrightarrow O_3 + M$$

Ozone is destroyed when UV light splits the molecule into an excited oxygen molecule and an excited oxygen atom. Ozone can also react with an oxygen atom to produce two oxygen molecules.

$$O_3 + h\nu(\lambda < 320 \text{ nm}) \longrightarrow O_2^* + O^*$$
$$O_3 + O \longrightarrow 2O_2$$

The balance between these reactions is dependant on the concentration of oxygen molecules (in the upper stratosphere this becomes too low) and the intensity of the UV radiation (in the troposhere this becomes too low), so the peak of ozone formation is in the lower statosphere. Taken together these reactions result in an equilbrium concentration of ozone in the ozone layer and the removal of much of the UV part of the solar spectrum before it reaches the Earth's surface (Figure 50.2).

These reactions alone do not account entirely for the observed ozone levels in the stratosphere; there are a number of others leading to the destruction of ozone including its reaction with radicals such as ·OH, ·NO, ·Cl and ·ClO. In recent years the chlorine and nitrogen species have inceased in concentration due to human activities, leading to concerns over the destruction of the ozone layer.

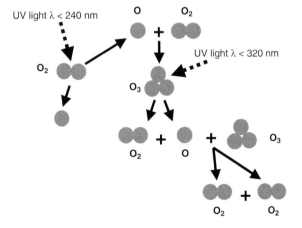

Figure 50.2 Formation and destruction of ozone.

The UV Shield

Although UV light was essential to the original evolution of life in the oceans, it is highly damaging to living cells. The UV C range is effectively absorbed by the atmosphere before reaching the Earth's surface, and although the UV B and UV A ranges are greatly reduced in intensity some reaches the ground (Table 50.1; Figure 50.3). With a decline in the ozone concentration more UV light reaches the Earth's surface.

Topic 50 The ozone layer

Table 50.1 Components of ultraviolet light.

	Wavelength (nm)	% of solar flux	Effects
UV A	315–400	7	The least damaging
UV B	280–315	1.5	Prolonged exposure harmful to plants and animals
UV C	<280	0.5	Rapidly damages/kills living cells

Exposure to UV light results in damage to important biochemical molecules, such as proteins and DNA. In humans, this can result in tanning, sunburn, aging of the skin, cataracts and skin cancer (basal cell carcinoma and the more dangerous malignant melanoma). Estimates vary, but it has been suggested that a 1% decrease in the ozone concentration leads to a 2% increase in UV B intensity and a 4% increase in skin cancer. Higher UV A and UV B levels could also affect the photosynthetic efficiency of plants, leading to decreased crop production and decreased primary productivity in natural ecosystems.

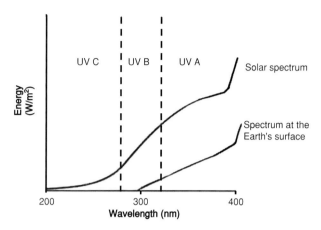

Figure 50.3 The effect of the UV shield on incoming solar radiation.

PHOTOCHEMICAL REACTIONS

Most chemical reactions are considered to occur when molecules collide and have sufficient energy to break chemical bonds and form new ones. Photochemical reactions occur when a molecule, atom or ion absorbs a quantum of light energy. The energy absorbed is equal to $h\nu$ (Planck's constant × the frequency of the light), which raises the molecule to an excited, and therefore reactive, state.

$$AB + h\nu \longrightarrow AB^*$$

The term $h\nu$ indicates the light energy absorbed and $*$ signifies the excited state. The excited molecule may then decompose or react with other molecules:

$$AB^* \longrightarrow A + B$$
$$AB^* + \text{reactants} \longrightarrow \text{products}$$

Although some photochemical reactions are initiated by visible light, it is more common for them to be initiated by UV ($\lambda < 400$ nm) light. Because there is an inverse relationship between the wavelength of light and energy, reactions requiring greater energy will require light of shorter wavelengths; longer wavelengths will not provide sufficient energy for the photochemical reaction. For example, photolysis of the oxygen molecule requires light of shorter wavelength (higher energy) than photolysis of ozone.

FREE RADICALS

Free radicals are molecules or ions with one or more unpaired electrons and which are therefore generally very reactive, often having only a transitory existence. The unpaired electron is commonly denoted as · beside or above the radical species.

The neutral hydroxyl radical, ·OH, is highly reactive and short-lived, and an important intermediary in many atmospheric chemistry reactions.

Nitric oxide (NO), nitrogen dioxide (NO_2), and molecular oxygen (O_2), all have unpaired electrons, but are much more stable; molecular oxygen is rarely written as a free radical.

51. Damage to the Ozone Layer

In recent years there has been a general decline of the order of 0.5% per year in measured stratospheric ozone levels. Considerable concern has been expressed over this trend and the annual formation of the more localised Antarctic ozone hole each spring in the southern hemisphere, where total atmospheric ozone falls to below 50% of the normal value, with almost complete loss of ozone at 15–20 km altitude. As described in Topic 50, the resultant increase in exposure to UV light has implications for human health, plant and animal growth, food production and the productivity of natural ecosystems.

Catalytic Ozone Destruction

Some naturally produced molecules such as methyl chloride and nitrous oxide, as well as a number of anthropogenically produced pollutant gases, are able to destroy stratospheric ozone molecules in a catalytic manner.

Chlorine-Containing Species

Most atmospheric chlorine derived from natural sources is present as HCl gas, from combustion or volcanic activity, or as sea spray. As this chloride is in the ionic form it is rapidly washed out of the troposphere in rainfall. Methyl chloride from natural sources, and molecules such as the chlorofluorocarbons (CFCs) and related compounds, are very unreactive in the troposphere, allowing them to persist long enough to migrate into the stratosphere. In the higher intensity UV of the stratosphere the molecules undergo photolytic decomposition, producing chlorine radicals which cause catalytic destruction of ozone molecules.

Chlorofluorocarbons undergo photolytic decomposition to produce a chlorine radical:

$$CFCl_3 + h\nu(\lambda < 290 \text{ nm}) \longrightarrow \cdot CFCl_2 + \cdot Cl$$

The residual CFC radical can generate additional chlorine radicals. The chlorine radical can then react with an ozone molecule to produce a chlorine monoxide radical, which can in turn react with an oxygen atom regenerating the chlorine radical:

$$\cdot Cl + O_3 \longrightarrow \cdot ClO + O_2$$
$$\cdot ClO + O \longrightarrow \cdot Cl + O_2$$

The overall effect of these two reactions is:

$$O_3 + O \xrightarrow{\cdot Cl} 2O_2$$

The chlorine radical is acting as a catalyst in the destruction of ozone; consequently a single chlorine radical can destroy many ozone molecules before it is consumed in one of a number of other reactions incorporating the chlorine atoms into less reactive reservoir species such as hydrogen chloride and chlorine nitrate. Bromine containing molecules behave in a similar manner, but are even more destructive than their chlorine analogues.

Nitrous Oxide

The NO_x gases nitric oxide (NO) and nitrogen dioxide (NO_2) are water soluble and reactive in the troposphere; they are rapidly washed out in rainfall, so do not persist long enough to reach the stratosphere. By contrast, nitrous oxide (N_2O) is not very soluble in water, is not a radical and does not absorb visible light and photolyse, so is unreactive in the troposphere and able to migrate up to the stratosphere. In the stratosphere, nitrous oxide can react with excited state oxygen atoms to produce nitric oxide:

$$N_2O + O^* \longrightarrow 2NO$$

Topic 51 Damage to the ozone layer

Nitric oxide is a radical and once formed can catalyse ozone destruction in a similar way to the chlorine radical. The nitrous oxide reacts with an ozone molecule to produce nitrogen dioxide, which can in turn react with an oxygen atom, regenerating the nitrous oxide:

$$NO + O_3 \longrightarrow NO_2 + O_2$$
$$NO_2 + O \longrightarrow NO + O_2$$

The overall effect of these two reactions is:

$$O_3 + O \xrightarrow{NO} 2O_2$$

As with the chlorine radical, nitric oxide is acting as a catalyst in the destruction of ozone molecules. Although there are other mechanisms to produce nitric oxide in the stratosphere, nitrous oxide is the main source.

Sources of CFCs and Related Compounds

CFCs are molecules containing only carbon, fluorine and chlorine. They are a class of synthetic compounds first developed in the 1930s. They have a number of properties, including low boiling point, non-toxic, non-flammable and chemically inert, which makes them ideal for applications such as blowing foams, as cleaning solvents for electronic circuits, as refrigerant gases, and as propellant gases in aerosol sprays. The halons are bromine-containing fluorocarbons used in fire extinguishers, and methyl bromide is used as a soil fumigant. World production of CFCs rose to approximately 800 kilotonne per year in the 1980s, a large proportion of which was lost to the atmosphere, resulting in a peak tropospheric concentration of just under 4 ppbv chlorine in the late 1990s.

The ozone depleting potential (ODP) is calculated as the expected long term impact on ozone compared to CFC11 which is given the value 1.0 (Table 51.1). This takes into account the atmospheric lifetime, reactivity and chlorine content of the molecule. CFC11, CFC12 and CFC113 taken together are considered to have been responsible for 80% of the ozone loss to date.

Table 51.1 Ozone depleting potentials of some CFCs, halons, HCFCs and HFCs.

	Formula	ODP		Formula	ODP
CFC11	$CFCl_3$	1.0	HCFC22	$CHClF_2$	0.05
CFC12	CF_2Cl_2	1.0	HCFC123	$CHCl_2CF_3$	0.02
CFC113	$CF_2ClCFCl_2$	0.8	HCFC141b	CH_3CCl_2F	0.1
CFC114	CF_2ClCF_2Cl	1.0	HCFC142b	CH_3CClF_2	0.07
CFC115	CF_2ClCF_3	0.6	HFC125	CHF_2CF_3	0
Halon 1301	CF_3Br	10.0	HFC134a	CH_2FCF_3	0
Halon 1211	CF_2ClBr	3.0	HFC143a	CH_3CF_3	0
Methyl bromide	CH_3Br	0.6	HFC152a	CHF_2CF_3	0

Since the effects of CFC on the ozone layer and the Antarctic ozone hole in particular have become apparent, alternative molecules with lower ozone depletion potentials have been increasingly substituted for the CFCs. The hydrochlorofluorocarbons (HCFCs) contain at least one hydrogen atom in the molecule. Replacing some of the chlorine in the molecule with hydrogen atoms makes the molecule more reactive, reducing the atmospheric lifetime and the ODP, but increasing flammability, hence making HCFCs unsuitable for some applications. Increasing the fluorine content of the molecule makes it more stable; the hydrofluorocarbons, which contain no chlorine atoms, consequently have zero ozone depleting potential. Unfortunately, they are good absorbers of IR radiation and consequently have high global warming potential (see Topic 49).

Sources of Nitrous Oxide

Nitrous oxide is formed naturally during denitrification in anaerobic environments and to a small extent during nitrification in aerobic soils (see Topics 12 and 32). The increased use of nitrogen fertilisers in agriculture has meant that there is a greater potential for dentrification when some of that nitrogen is leached into anaerobic environments. Tropospheric nitrous oxide levels have risen by about 15% over pre-industrial levels.

The Antarctic Ozone Hole

In the early 1980s it became clear that a major loss of stratospheric ozone was occurring over Antarctica each September and October, with almost complete loss of ozone at altitudes between 15 and 20 km. It also soon became clear that the degree of ozone loss was becoming progressively greater.

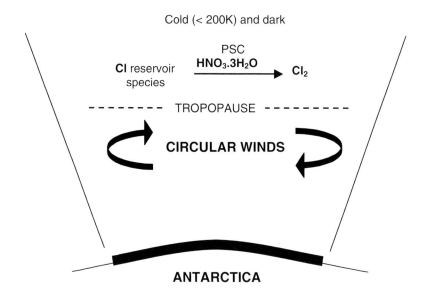

Figure 51.1 The polar vortex in winter.

In the Antarctic, and to a lesser extent in the Arctic, conditions develop in the spring where very rapid loss of stratospheric ozone can occur. During the long dark Antarctic winter a circular wind pattern is set up due to the intense cold and the Earth's rotation, the polar vortex (Figure 51.1). The vortex is largely distinct from the air mass at lower latitudes, allowing the chemical reactions to occur in isolation. When temperatures in the lower stratosphere fall below 200 K, polar stratospheric clouds (PSC) form. Type 1 PSC consist of 1 μm diameter nitric acid: water (1:3) ice and type 2 PSC consist of 10 μm diameter pure water ice. These act as surfaces on which the accumulated chlorine and nitrogen reservoir species (mainly hydrogen chloride and chlorine nitrate) react to form chlorine and hypochlorous acid:

$$HCl + ClONO_2 \longrightarrow Cl_2 + HNO_3$$
$$ClONO_2 + H_2O \longrightarrow HOCl + HNO_3$$

In the absence of sunlight no photochemical reactions can occur, but with the onset of the Antarctic spring and the return of sunlight, photochemical decomposition of these more reactive chlorine species forms chlorine radicals:

$$Cl_2 + h\nu \longrightarrow 2 \cdot Cl$$
$$HOCl + h\nu \longrightarrow \cdot Cl + \cdot OH$$

These participate in rapid catalytic destruction of ozone, leading to almost total loss of ozone in the lower stratosphere within a few days and 50% loss of the total ozone in the atmospheric column. As temperatures rise in the late spring the air circulation changes and the vortex breaks up. Mixing of air masses leads to polar ozone levels returning to normal, but a general reduction in ozone levels at lower latitudes exposing populations in Australia and South America to elevated levels of UV B. In addition, phytoplankton productivity in the Antarctic ocean, which forms the basis of the food chain, is likely to be reduced.

The Arctic is slightly less cold than the Antarctic, the polar vortex is less pronounced and less persistent; therefore conditions for conversion of the chlorine reservoir species to chlorine as a prelude to catastrophic ozone destruction are less favourable. There have, however, been years such as 1995 when there has been significant loss of ozone. The formation of an Arctic ozone hole is potentially more serious because of the larger human populations living close to the Arctic circle.

The Montreal Protocol

The Montreal protocol, signed in 1987, and the several revisions in later years have marked a successful world response to ozone depletion. The initial protocol sought to reduce the atmospheric chlorine loading to 2 ppbv by reducing CFC use and substitution with HCFCs. Revisions brought in total bans on many ozone depleting substances (CFCs, HCFCs, halons and methyl bromide, and a number of chlorinated solvents such as carbon tetrachloride and methyl chloroform), and subsequent revisions brought forward the phase-out dates. The success of the protocol can be seen in that atmospheric chlorine loading is now declining, but many of the CFCs are so long lived that the 2 ppbv chlorine target is unlikely to be achieved before 2050.

52. Acid Rain

As shown in Topic 11, unpolluted rainwater has a pH of 5.7 because of dissolution of atmospheric CO_2 and buffering by the carbonate system. Other gases, mostly of anthropogenic origin, are present in the atmosphere and dissolve in rainwater to produce stronger acids. This leads to rainwater with a pH below 5, which is called 'acid rain', but this term can also include dry deposition of particulates with acid properties. Three environmental consequences of acid rain have been identified:

(1) Acidification of surface waters (rivers and lakes)
(2) Direct damage to vegetation (especially in forests)
(3) Direct damage to buildings and other structures (especially marble/limestone and metal).

Acid Gases: Sources and Chemistry

The main acid producing gases are sulfur dioxide (SO_2) and nitrogen oxides, especially nitrogen dioxide (NO_2) and nitric oxide (NO), which are collectively known as NO_x. These result in the formation of sulfuric and nitric acids. Ammonia (NH_3) is also an important contributor to acidification because, although it is itself an alkaline gas, it dissolves in water to form ammonium ions (NH_4^+). These undergo nitrification in aerobic environments (see Topic 32), which is an acidifying process.

The major source of SO_2 is burning of fossil fuels in electricity generating power stations, with combustion of fossil fuels by industry being a small, but significant, input. Most NO_x is produced by fossil fuel combustion by cars and lorries, with power production being a smaller input. The main source of ammonia is agriculture, especially animal wastes. Table 52.1 shows the emissions of these gases in the UK in 2002.

Table 52.1 Emissions of acid producing gases in the UK in 2002.

	Total emissions (kt)	% of emissions from power stations	% of emissions from road transport	% of emissions from agriculture
NO_x	1582	24	45	–
SO_2	1002	68	–	–
NH_3	285	–	4	89 (75% animal wastes)

Source: http://www.naei.org.uk/ (UK National Atmospheric Emissions Inventory).

The chemistry of SO_2 and NO_x in the atmosphere is highly complex and can include both homogeneous (entirely gas phase) reactions as well as heterogeneous (gas–liquid phase) reactions. There may also be involvement of free radical species (uncharged atoms or molecules that have an unpaired electron, and which are highly reactive), and of oxidising species such as ozone (O_3) and hydrogen peroxide (H_2O_2). Photochemical reactions may occur in the gas phase, catalysed by ultraviolet (UV) light.

In the gas phase (homogeneous reaction) SO_2 can react with a hydroxyl free radical (·OH), which is produced in the atmosphere by the photolysis of ozone.

$$O_3 \xrightarrow{UV\ light} O_2 + O^* \xrightarrow{+H_2O} 2\cdot OH$$
$$SO_2 + \cdot OH \longrightarrow \cdot HSO_3$$
$$\cdot HSO_3 + O_2 \longrightarrow \cdot HO_2 + SO_3$$
$$SO_3 + H_2O \longrightarrow H_2SO_4$$

The ·HO_2 (hydroperoxyl) radical can react with NO to produce nitrogen dioxide, which dissolves to form nitric acid:

$$\cdot HO_2 + NO \longrightarrow NO_2 + \cdot OH \longrightarrow HNO_3$$

Alternatively, SO_2 can dissolve in water (heterogeneous reaction):

$$SO_{2(g)} + H_2O \longrightarrow SO_{2(aq)}$$
$$SO_{2(aq)} + H_2O \longrightarrow H_2SO_3$$
$$2H_2SO_3 + O_2 \longrightarrow 2H_2SO_4$$

Nitrogen dioxide can react with ozone to form nitric acid:

$$NO_2 + O_3 \longrightarrow NO_3 + O_2$$
$$NO_3 + NO_2 \longrightarrow N_2O_5$$
$$N_2O_5 + H_2O \longrightarrow 2HNO_3$$

Impact of Acid Rain on Soils and Surface Waters

Areas that are particularly susceptible to the effects of acid rain are those underlain by acidic rocks, such as granite and sandstone, and which lie downwind of major industrial centres where most of the SO_2 and NO_x is produced. The acid gases persist in the atmosphere for 1–2 weeks, which allows dispersal up to a few hundred kilometres. Thus acid rain falling in the UK has originated from gases produced in central Europe, while the gases produced in the UK result in acid rain in Scandanavia. Similarly, in the USA, the area most affected by acid rain is in the northeast, upwind of the major industrial centres.

The soils formed over these types of rock are themselves acidic and have a poor buffering capacity (see Topics 11 and 13). The input of acid rain overcomes what buffering capacity there is, and the soil pH falls. As sulfate and nitrate ions are poorly held in soil (see Topics 8, 32 and 33) they readily leach, taking cations with them to maintain charge balance. This results initially in loss of exchangeable cations such as Ca^{2+} (see Topic 7), which is also an acidifying process. Aluminium becomes soluble in significant amounts when the soil pH falls below about 4.5 (see Topic 9) and so becomes a major, highly toxic, cation in the leachate, killing off fish, macrophytes and other aquatic organisms in lakes and rivers.

Reduction in Emissions of Acid Producing Gases

Because of the transnational nature of acid rain, controls on emissions of the acid producing gases have been set by international agreements. Under the Gothenburg protocol of the United Nations Economic Commission for Europe (UNECE), published in 1999, ceilings were set for SO_2 and NO_x to cut emissions from 1990 levels by at least 63% and 41%, respectively, by 2010. It is estimated that this will reduce the area in Europe affected by acidification from the 1990 figure of 93 million hectares to 15 million hectares in 2010. The UK targets were set at 625 kilotonnes SO_2 and 1181 kilotonnes NO_x per year, which have been reduced further to 585 kt and 1167 kt by the UK government. The progress in meeting these targets, and the overall pattern of emissions since 1970, can be seen in Figure 52.1

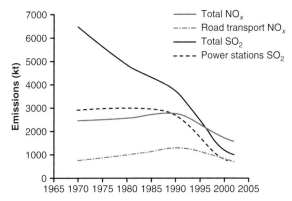

Figure 52.1 Reductions in SO_2 and NO_x emissions in the UK since 1970.
Source: www.naei.org.uk (UK National Atmospheric Emissions Inventory).

Total SO_2 emissions have declined steadily from 1970, but up to 1990 this was due mainly to the reduction in domestic coal burning and in lower inputs from industry. Emissions from the main source, electricity producing power stations, were reasonably steady during this period, with only a slight reduction due to changes in the type of coal used. Since 1990, the rate of reduction in emissions of SO_2 from power stations has increased because of the greater use of gas as a fuel, greater use of low sulfur coal, the relatively greater importance of nuclear energy, and the introduction of technology to scrub SO_2 out of the emissions from power stations. The latter include flue gas desulfurisation, which has been fitted at two of the largest power stations in the UK, Drax and Ratcliffe. For this, the flue gases are passed through a hydrated lime or limestone slurry, which reacts with sulfur dioxide to produce calcium sulfite:

$$\underset{\text{hydrated lime}}{Ca(OH)_2} + SO_2 \longrightarrow CaSO_3 + H_2O$$

$$\underset{\text{limestone}}{CaCO_3} + SO_2 \longrightarrow CaSO_3 + CO_2$$

The calcium sulfite is oxidised to gypsum (calcium sulfate):

$$2CaSO_3 + O_2 + 2H_2O \longrightarrow 2CaSO_4.2H_2O$$

The changes in NO_x followed a different pattern. Between 1970 and 1990, both total NO_x and road transport NO_x emissions rose slightly, due to the increase in traffic during this period. A decline began in the early 1990s as three-way catalytic converters started to be fitted to new cars.

Index

Note, entries are indexed by topic number.

acid mine drainage, 9, 46
acid producing gases, 52
acid rain, 9, 52
actinomycetes, 26
activated sludge process, 41
adsorption, 8, 34
aeration of sediments, 30
aeration of soils, 30
aeration of water bodies, 30
aerobic respiration, 12, 25, 31
albedo, 28
alfisols, 15
algal bloom, 42
aluminium oxides, 4, 34
ammonia volatilisation, 32
anaerobic respiration, 12
Antarctic ozone hole, 51
aquifer, 21, 40
archaebacteria, 26
arsenic, 45
atmosphere of early earth, 23
atmospheric boundary layer, 23
authigenic sediments, 18
autochthonous population, 31
autotroph, 25
autotrophic bacteria, 13, 25
available water capacity, 29

back radiation, 28
base saturation, 7, 15
bioaccumulation, 37
biogenic sediments, 18
biological weathering, 2
birnessite, 4
blue-green algae, 26
BOD, 41, 42
brown earth soils, 15
BTEX, 44
buffer capacity, 11
buffer curves, 11

cadmium, 45
capillary water, 29
carbon cycle, 31
carbonate equilibria, 11, 31
cation exchange capacity, 3, 7
CFCs, 49, 51
chelation, 2, 10

chemical composition of atmosphere, 23
chemical composition of earth's crust, 1
chemical composition of ground water, 21
chemical composition of rain water, 21
chemical composition of sea water, 22
chemical composition of soil atmosphere, 30
chemical weathering, 2
chemisorption, 8, 34
chemoautotrophic bacteria, 11, 25, 26, 32, 33, 39
Chernobyl, 7
chlorite, 3
chromium, 45
clay domain, 6, 17
clay minerals, 3, 7
CLEA model, 43
condensation cell, 24
copper, 8, 10, 45
Coriolis force, 24
cosmic radiation, 47
cosmogenic radionuclides, 47
cyanophytes, 26

DDT, 37
denitrification, 12, 32, 49, 51
detrital sediments, 18
diagenesis, 18
diffuse sources of pollution, 36
diffusion of gases, 30
diffusion of nutrient ions, 27
dioxin, 44
DNAPL, 44
drinking water, 40

Eh, 12
element balance of oceans, 22
energy balance at earth's surface, 28, 49
eubacteria, 26
eutrophication, 42
evapotranspiration, 20, 27, 39
exchangeable acidity, 11
exchangeable cations, 7, 27, 35

fermentation, 12
ferrihydrite, 4
fertiliser response curve, 38
fertilisers, 38, 39
field capacity, 29, 39
fixation of ions, 7, 35
flue gas desulphurisation, 52
free radicals, 50
fulvic acid, 5
fungi, 26
furan, 44

gibbsite, 4
gley soils, 15
global energy balance, 24
goethite, 4
gravitational water, 29
greenhouse gases, 49
ground water gley soils, 15
groundwater, 21, 40

Hadley cell, 24
haematite, 4
half life, radioactive, 47
heterotrophic bacteria, 13, 25, 26, 42
histosols, 15
Hofmeister series, 6, 7
humic acid, 5
humification, 5
humified organic matter, 5, 6, 7, 10, 31, 32, 33, 34
humin, 5
hydrated radius of ions, 6, 7
hydrological cycle, 20, 29, 39
hydrous oxides, 4, 6, 7, 8
hygoscopic water, 29

illite, 3, 35
inositol hexaphosphate, 34
intertropical convergence zone, 24
ion exchange, 7, 35
iron oxides, 4, 34
iron oxidising bacteria, 11
iron reduction, 12, 18
isomorphous substitution, 1, 3

kaolinite, 3
Kyoto protocol, 49

Index

Langmuir equation, 8
leaching, 7, 39
lead, 45
lepidocrocite, 4
lichens, 26
ligand exchange, 8
LNAPL, 44

macrofauna, 13, 26
macronutrients, 27
manganese oxides, 4
manganese reduction, 12, 18
mass flow of nutrients, 27
mercury, 45
methaemoglobinaemia, 40
methane, 12, 49
microaggregates, 17
microfauna, 13, 26
micronutrients, 27, 45
mine tailings, 46
mineral weathering, 2, 27, 34, 35
mineralisation/immobilisation, 32, 33, 34
mining spoil, 46
Montreal protocol, 51
mor humus, 15
mull humus, 15
mycorrhiza, 26

Nernst equation, 12
nickel, 45
nitrate directive, 40
nitrate toxicity, 40
nitrate vulnerable zones, 40
nitrification/nitrifying bacteria, 11, 25, 32, 39
nitrogen, 32, 38, 39, 40, 50, 51, 52
nitrogen fixation, 32
non-exchangeable acidity, 11
non-silicate minerals, 1
nutrient availability, 27
nutrient transport, 27

ocean currents, 22
octanol-water partition coefficient, 44
organic matter decomposition, 5, 31
oxygen sag curve, 36, 42
ozone layer depletion, 51

PAH, 44
particle size analysis of soil, 16
PCB, 44
PDBE, 44
peat soils, 15
peds, 17
permanent charge, 3
permanent wilting point, 29
pH, 11
pH buffering, 11
pH dependant charge, 4, 5
phosphate solution equilibria, 34
phosphorus, 8, 34, 38
photoautotrophs, 25
photochemical reactions, 50, 52
photosynthesis, 25, 31
physical weathering, 2
plant nutrients, 27
PM10, 44
podzol soils, 15
point of net zero charge, 4, 6
point sources of pollution, 36
pollutant linkage model, 43
potassium, 35, 38
potassium fixation, 35
potentially toxic elements (PTEs), 10, 45
primary minerals, 1
primary productivity, 25
primordial radionuclides, 47
protozoa, 26

radioactive decay, 47, 48
radioactivity, exposure, 48
radioactivity, health effects, 48
radiocarbon dating, 5, 48
radiochemical dating techniques, 48
radon, 48
redox potential, 12, 18
relative humidity, 20
residence times, hydrosphere, 20
river catchment, 21
root systems, 27

salinity of sea water, 22
saturation vapour pressure, 20
secondary minerals, 1, 3, 4
sediment transport, 19, 22
selenium, 45
sewage sludge, 41, 42
silicate minerals, 1, 3
smectite, 3
soil aggregation, 17
soil atmosphere, 30
soil forming factors, 13, 14
soil guideline values, 43, 45
soil horizons, 14
soil moisture deficit, 29
soil parent materials, 13
soil pores, 17, 29, 30
soil profiles, 14
soil water, 29
soil water potential, 29
solubility product, 9
source-pathway-receptor model, 43
spodosols, 15
Stokes' law, 16
stomach cancer, 40
storm hydrograph, 21
stratosphere, 23, 50
sulfate reduction, 12, 18, 33
sulfur cycle, 33, 52
sulfur oxidising bacteria, 11, 25, 33
sulfur reducing bacteria, 12, 33
surface water gley soils, 15

temperature cycles in soil profiles, 28
thermal stratification of lakes, 21, 28, 30
thermal stratification of oceans, 28
thermohaline circulation, 22
tortuosity of pore system, 27, 30
trace elements, 10, 27, 45
troposphere, 23
tropospheric ozone, 52

UV radiation, 50, 51

variable charge surfaces, 4, 5
vermiculite, 3

water, physical properties, 20

zinc, 45
zymogenous population, 31